The World Wide Wi-Fi

The World Wide Wi-Fi

Technological Trends and Business Strategies

TEIK-KHEONG (TK) TAN
BENNY BING

WILEY-INTERSCIENCE

A JOHN WILEY & SONS, INC., PUBLICATION

Library of Congress Cataloging-in-Publication Data:

Bing, Benny.
   The world wide Wi-Fi : technological trends and business strategies /
Benny Bing & TK Tan.
      p. cm.
"A Wiley-Interscience Publication."
Includes bibliographical references and index.
      ISBN 0-471-46356-6 (Cloth)
   1. Wireless LANs.   2. IEEE 801.11 (Standard)   3. Local area networks
industry.   I. Tan, TK, 1964–   II. Title.
   TK5105.78.B56 2003
   004.6'8—dc21   2003011396

Printed in the United States of America.

10 9 8 7 6 5 4 3 2 1

For her dedication and courage and for teaching me the true meaning of life, I dedicate this book to my beloved mother, Sow-Lan.

- Teik-Kheong (TK) Tan

To my mum, as always.

- Benny Bing

Contents

Chapter 3 Wi-Fi Network Security   **41**

Foreword

Over the past few years, wireless LANs have played a key role in revolutionizing the use of technology in our society. Today Wi-Fi is fast becoming commonplace in enterprise systems, education, manufacturing, healthcare, retail, logistics, and distribution infrastructures and most recently in the home. In short, wireless connectivity is permeating every aspect of our lives.

What makes the IEEE 802.11 or Wi-Fi so successful? Productivity increases are driving businesses to install wireless LANs. Demand for the technology has surged recently on a global basis. The expected boost in worker output helps offset possible security and performance failings. The good news is that wireless LAN standards development is progressing at a steady pace to address many of the outstanding issues. Judging from the rate of market entry for Wi-Fi products today, it is highly possible that we will have solutions to security, quality of service, and higher throughput faster than originally planned.

In the 21st century, we are seeing new wireless possibilities emerging for greater system flexibility, higher speeds, better security, more robust quality of service, and regular use of wireless in the home and business. The need to understand the business implications of wireless LANs has never been greater today. For that reason alone, this book has certainly delivered on its promise as advertised.

Like most technology books, this book provides good technical coverage on the many aspects of wireless LANs. What sets this book apart from its competition are the business implications of the technology. The authors illustrate this strategic linkage between technology and how it influences business decisions in Chapters 6 and 7 effectively. Chapter 8 provides a future roadmap of wireless LANs for both the technologist and business investor. The inclusion of IEEE standards, security, quality of service, and

Wi-Fi hotspots make this book an interesting read. The authors have adequately highlighted the key practical issues of wireless LANs. The caliber of the authors is of the first ranks in my opinion. TK Tan and Benny are clearly subject matter experts in this field and they bring to bear their vast experience in writing this book.

Overall, *World Wide Wi-Fi: Technological Trends and Business Strategies* is an excellent guide for any network professional or business executive who needs to understand the business implications of wireless LANs. It has all the necessary information for decision making pertaining to investment and understanding of wireless LANs today and tomorrow. It explains every aspect simply and straight to the point.

Stuart J. Kerry
Chairman, IEEE 802.11 Working Group
Business Executive, Philips Semiconductors

Preface

IEEE 802.11 wireless local area networks (LANs) are becoming ubiquitous and increasingly relied upon. From airport lounges and hotel meeting rooms to cafés and restaurants across the globe, these networks are being built for mobile professionals to stay connected to the Internet.

The demand for wireless access to LANs is fueled by the growth of mobile computing devices such as laptops, personal digital assistants, and pocket PCs, and the desire of mobile professionals for continual connections to the network without having to "plug in". There will be over a billion mobile devices by 2003. 802.11b, or Wi-Fi, dominates wireless networking in the business world, and it has also invaded the home market. According to a recent study by IDC, wireless LAN equipment revenue will reach $3.2 billion in 2005. Over ninety percent of that is represented by Wi-Fi. However, there is much confusion in the 802.11 wireless LAN industry today. Vendors and end-users are faced with the difficult task of deciding which version of 802.11 to adopt. The emergence of combo solutions such as the a/g and a/b combinations of the 802.11 standards has further clouded the decision-making process.

This book is designed for networking professionals and IT executives who are interested in developing tactics and strategies that take maximum advantage of the exciting Wi-Fi wireless LAN market. The technology and business implications are presented in an easy-to-understand manner. Specifically, it covers important wireless LAN design and deployment issues, as well as market dynamics, market segmentation, service provider, enterprise, and chipset strategies. The value of this book lies in its analysis of the current wireless LAN market as well as future threats to and opportunities for the stakeholders of this industry. Both the technical and market insights will enable the reader to look ahead and be positioned to profit from this competitive industry.

The book begins with an overview of the technology and the market segments followed by an examination of the key drivers behind the success of the 802.11b technology. Emerging Wi-Fi standards like 802.11g and its impact on all market segments are then covered. Case studies involving enterprise, home, and public access wireless LAN deployment as well as the implications of standards development will also be analyzed extensively. Finally, the inter relationship of the 802.11a, b, and g standards will be presented.

We hope we have succeeded in achieving our objectives in writing this book. A Web-based resource for the book is planned. Feedback from readers on how the book can be improved and the topics they would like to see in future editions are most welcome and can be directed to the authors at tktan@ieee. org and bennybing@ieee.org.

Teik-Kheong (TK) Tan
Benny Bing
June 2003

Acknowledgments

Writing a book on a topic such as wireless LANs has its challenges. The dynamic market landscape constantly floods us with new information. Thanks to the generous support of friends and family, this book provides a snapshot of today's technologies and a roadmap for future wireless LAN technologies.

Special thanks are due to our friend and colleague, Stuart Kerry, for writing the Foreword and his constant encouragement. We are also grateful to Val Moliere and Kirsten Rohstedt of Wiley-Interscience for their support, patience, and helpful suggestions throughout the writing process.

Chapter 1

Introduction

Wireless LANs became part of the wireless revolution at the turn of the millennium, imitating the success of cellphones in the prior decade. Such networks combine the power of wireless access with mobile computing, delivering high data rates on the unlicensed radio spectrum. In addition, the same high-speed wireless LAN cards can be used virtually anywhere, from the office and public spaces to the home. The cards can be used in a laptop, personal digital assistant (PDA), or Pocket PC and are typically available for less than $100 from a wide variety of electronics vendors, including low-priced makers whose products are stocked on retail shelves. The fact that large retail outlets such as Starbucks, GAP, and Sears are deploying 802.11b or Wi-Fi wireless LANs shows how prevalent the technology has now become.

The increasing popularity of Wi-Fi is seen as a rare bright spot for the communications industry. Currently, the annual revenue has exceeded US$1.6 billion and an estimated 25 million Wi-Fi-enabled computers and other personal computing devices are already in use in the U.S. and overseas. We are also starting to witness the exciting convergence of wireless communications and computing. Intel plans to incorporate Wi-Fi technology in all of the microprocessor chips it ships in 2003, providing tens of millions of desktop, laptop, and hand-held computing devices with built-in broadband wireless access [8]. These developments rival the popularity of wired Ethernet networks.

When wireless LANs were first deployed, they gave laptop and PDA users the same freedom with data that cellphones provided for voice. However, a wireless LAN need not transfer purely data traffic. It can also support packetized voice and video transmission. People today are spending huge amounts of money, even from office to office, calling by cellphones. With a wireless LAN infrastructure, it costs them a fraction of what it costs them using cellphones or any other equipment. Thus, voice telephony products based on wireless LAN standards have recently emerged. A more compelling use of wireless LANs is in overcoming the inherent limitations of wireless wide area networks (WANs). Current third-generation (3G) mobile telephony data rates have the potential to increase up to 2 Mbit/s, whereas wireless LANs already offer data rates of up to 54 Mbit/s and, unlike 3G, operate on unlicensed frequency bands. This has led some technologists to predict that eventually we are more likely to see dense urban broadband wireless LANs that are linked together into one network rather than widespread use of high-powered WAN handsets cramming many bits into expensive and narrow slices of radio spectrum.

1.1   Past Wireless Lessons

In the last few years, several mobile wireless technologies have fallen short of expectations and none have emerged to be the mainstream wireless technology even though they started with much promise and were put together by a huge group of multinational companies. Some prominent examples include Iridium's low-orbit satellites that allow global telephony, and the wireless application protocol (WAP) that permits Web surfing over mobile phones. Many of these technologies were aimed at extending the wireless revolution created by second-generation (2G) cellphones in the last decade.

When voice-centric 2G cellphones first became commercially available in the U.S. in 1983, an explosion of sales soon followed. Over 50 million subscribers were added in the first 15 years of this wireless revolution, which doubled every 18 months. Today, worldwide cellphone sales totaled over 400 million units annually. Interestingly, the current success model of wireless LANs is quite the reverse of that of 2G cellphones. 2G cellphones operate on a selected radio spectrum that is licensed to telecommunication carriers. Being expensive, the spectrum allocation is limited but is sufficient to support voice and low-speed Internet services. This is in contrast to wireless LANs that operate on larger chunks of spectrum that are freely usable by anybody without a license. The larger spectrum allocation allows wireless LANs to support a diverse range of multimedia and Internet services. The unlicensed nature removes exorbitant overheads associated with acquiring radio spectrum, lowering the barrier for new vendors, which increases competition and drives down product costs. This in turn provides a hotbed of entrepreneurial activity because it is the end-users (and not the carriers) who control and expand the network, even in remote, sparsely populated areas not economically serviceable by cellphone or cable networks.

The caveat for unlicensed operation is that user devices that join the network must first listen to the network to detect ongoing transmissions and defer transmission if the network is sensed to be busy. In addition, when the network is idle, users who wish to transmit must do so at the lowest power sufficient to maintain a connection of reasonable quality and at the desired range. Such etiquette practice does come with some useful device benefits, namely longer battery lifetime, lower prices, and interference avoidance.

Another key difference is that 2G cellphones employ dedicated circuit-switched connections for each individual user, which are maintained regardless of whether the user is inactive or transmitting. This in contrast to packet-switched wireless LANs that interleave packets from transmitting users over a single channel, thereby making it economical to implement what is termed "always on" communication. The concept of sharing network resources fits very nicely to users of unlicensed spectrum since radio bandwidth is consumed only by users with data to transmit, ultimately leading to far more efficient utilization of wireless spectrum.

1.2   What Are Wireless LANs?

Wireless LANs provide high-speed, cable-free access for computer-to-computer information transfer, typically within a building. They possess all the functionality of wired LANs, but without the physical constraints of the wire itself. The wireless nature inherently allows easy implementation of broadcast/multicast services. When used with portable computing devices (e.g., notebook computers), wireless LANs are also known as cordless LANs because this term emphasizes the elimination of both power cord and network cable.

A substantial portion of the cost of LAN deployment is in interconnecting end-user devices, which many networking experts acknowledge can sometimes exceed the cost of computer hardware and software. A wireless LAN removes the labor and material costs inherent in wiring. It also offers the flexibility to reconfigure or add more fixed clients to the network without much planning effort and the cost of recabling, thereby making future upgrades inexpensive and easy. The ability to add new mobile computing devices quickly is another main consideration for choosing a wireless LAN. These advantages, coupled with improved data rates, interoperability, and the proliferation of cheaper, smaller, and more powerful computing devices have accelerated the adoption of wireless LAN technologies in recent years.

1.3   The 802.11 Standards

The Institute of Electrical and Electronic Engineers (IEEE) 802.11 wireless LAN standard was developed by the 802 Local and Metropolitan Area Networks Standards Committee (LMSC) of the IEEE Computer Society. The development of the standard was heavily influenced by existing wireless LAN products already available in the market. It follows a development path similar to the IEEE 802.3 standard, which was initiated after Ethernet was commercialized.

The original 802.11 standard evolved from six draft versions and the final draft was approved on June 26, 1997. The standard allows multiple vendors to develop interoperable products in the globally available 2.4-GHz industrial, scientific, and medical (ISM) unlicensed band. The standard later became a joint ISO/IEC and IEEE standard. Two improved versions ratified in September 1999 run faster than the first: 802.11b allows data rates of up to 11 megabits per second (Mbit/s) in the 2.4-GHz band, whereas a device conforming to the 5-GHz 802.11a standard can transmit up to 54 Mbit/s. Most recently, 802.11g—essentially a 2.4-GHz version of 802.11a—has been ratified. These new developments have significant implications in that the standard is now more than capable of supporting streaming audio and video traffic needed for multimedia Internet applications. Note that the data rates are shared access and a single user will not see 11 Mbits/s or 54 Mbit/s unless there is only one user transmitting. Also, these data rates represent the raw wireless rates and not the actual throughputs that take into account the massive overheads in wireless transmission. Since it is the through-put that is experienced by the user, this is a key parameter in distinguishing the performance of one 802.11 card from another card that is manufactured by a different vendor.

The 802.11 standard specifies wireless connectivity for fixed, portable, and moving clients in a limited geographic area. Specifically, it defines an interface between a wireless client and an access point, as well as among wireless clients. As in any 802 LAN standard such as 802.3 (Ethernet) and 802.5 (Token Ring), the 802.11 standard specifies data rates of at least 1 Mbit/s and defines only the physical (PHY) and medium access control (MAC) layers, which correspond to the first two layers of the Open System Interconnect (OSI) network hierarchy. However, the 802.11 MAC layer also performs functions that are usually associated with higher-layer protocols (e.g., fragmentation, error recovery, mobility management, power conservation). These additional functions allow the 802.11 MAC layer to conceal the unique characteristics of the wireless PHY layer from higher layers. Given the number of 802.11 standards (802.11a, 802.11b, and 802.11g) operating on different frequency bands (2.4- and 5-GHz), it is unclear which technology will eventually prevail, or whether solutions involving combined standards are required.

1.4   The Wi-Fi Alliance

Unlike current Bluetooth devices, the same Wi-Fi card can be used to access Wi-Fi networks configured by different vendors as the mobile professional travels from an office workgroup to a café to a convention site, to the home. To achieve such mobility using a single card, small differences in the way vendors implement the standards must be removed so that interface cards from one vendor will work with the access point from another. A nonprofit organization called the Wi-Fi Alliance or WFA (formerly the Wireless Ethernet Compatibility Alliance or WECA) was formed in 1999 to certify the conformance and interoperability of 802.11 products and to promote Wi-Fi as a global wireless LAN standard. Wireless Fidelity or Wi-Fi is the shorthand for 802.11 compliance certification by the WFA, considered a seal of approval for 802.11 products. The alliance has also embarked on two other important initiatives—Wireless Internet Service Provider Roaming (WISPr) and Wi-Fi Protected Access (WPA). The WFA currently comprises over 180 members and these member companies offer over 740 Wi-Fi products. A Return-on-Investment (ROI) calculator tool for IT managers is freely downloadable from the WFA website.

1.5   Wireless Home and Community Networks

The strong emergence of wireless LANs in the last few years has influenced a growing demand for home wireless networks. An important goal of such networks is to enable two or more users to share a dial-up modem, cable modem, or digital subscriber line (DSL) adapter for accessing the Internet. This enables everyone at home to simultaneously surf different Web sites and access e-mail through a single Internet service account. Although such residential networks are typically deployed on a smaller scale than enterprise networks (an exception being community networks), servicing the wide variety of entertainment traffic poses a challenging problem. To do this, a dynamic resource allocation scheme is needed, which can assign bandwidth, channel, and power levels based on current interference, propagation, and traffic conditions. Defining quality of service (QoS) metrics is a key step in developing such an allocation scheme. The 802.11e Task Group produced a draft standard on QoS provisioning for 802.11 wireless LANs in November 2002. The recommended changes primarily affect the MAC layer, which is common to all 802.11 standards (i.e., 802.11a, b, and g).

Setting up a Wi-Fi wireless LAN within the home is relatively straightforward, requiring a wireless residential gateway (a hub that converts wired DSL or cable network traffic into wireless data packets) and one or more Wi-Fi devices. Stick the wireless gateway in the window, and one can work in the backyard or a nearby pool or park. If the neighbors are cooperative enough to allow their mini-networks to be accessed in this manner, then a community network can be formed quickly; in this case, a wireless LAN service that is directly managed by members of the community. Such "grassroots" Wi-Fi service providers (or "micro-operators" [6]) are a growing trend and have sprung to life in San Francisco, Seattle, New York, Canberra, and other major cities, offering free Internet access and community-service networking. Some notable examples of Wi-Fi community networks located in metropolitan areas include the Bay Area Wireless User Group (www.bawug.org), the New York City Wireless (www.nycwireless.net), and the London-based group Consume Net (www.consume.net) [4].

1.6   Public Wi-Fi Services

Public Wi-Fi services (or Wi-Fi hotspots) allow users to access corporate networks or the Internet at broadband speeds from airport lounges, hotels, or coffee houses. This can be done on a laptop without having to squint at small mobile-phone screens. For the U.S. market, this service has potential since, unlike the Japanese, where an estimated 72% of cellphone users routinely connect to the Internet, only a mere 6% of U.S. subscribers do so [10]. The International Data Corporation (IDC) projected that the public Wi-Fi service market will take off in the next few years, growing from the current 23,700 users to 609,200 users by 2004 and boosting revenues from $8.4 million to $204.7 million. Based on the most recent IDC data, the European hotspot market increased encouragingly by 327%, from 269 locations at the end of 2001 to around 1,150 locations at the end of 2002. These Wi-Fi hotspots will primarily target business travelers [9], as opposed to home networks, which are consumer-centric.

There are a handful of Wi-Fi providers that offer services in the public space. Aggregators like Boingo™ Wireless, GRIC Communications, and iPass do not own the access points or building infrastructure but strike deals with the local wireless providers. Boingo is teaming with FiberLink to offer business users secure virtual private network (VPN) service that includes integrated firewalls, antivirus software, and management support.

In late 2001, T-Mobile USA (then VoiceStream) acquired most of MobileStar's 802.11 access points in airports, hotel chains, 145 Borders Books and Music stores, and 1,200 Starbucks cafés. There are now plans to bolster the 2,000 access points deployed in 2002 with at least 5,000 by the end of 2003, including over 1,000 Kinko outlets. In a more recent development, Intel, At&T, and IBM launched Cometa Networks at the beginning of December 2002. The new company plans to roll out 20,000 access points across the 50 states of the U.S. by 2004.

Being layer-2 technologies, wireless LANs offer limited Internet roaming capabilities and global user management features, including billing and identification features commonly available in cellphone systems. Nokia recently proposed a system that efficiently combines wireless LAN access with the widely deployed Global System for Mobile Communications/General Packet Radio Service (GSM/GPRS) roaming infrastructure [3]. Moreover, the architecture exploits GSM authentication, SIM-based user management, and secured billing features. This gives the cellphone operator a major competitive advantage over Internet service providers, including current hotspot providers, who have neither a large mobile customer base nor a cellphone-type roaming service. In mid-March 2003, Boingo announced a partnership with T-Mobile to allow users to connect to 802.11 hotspots and GPRS services using the same account and software.

In early 2001, the WFA initiated the Wireless Internet Service Roaming (WISPr) project to allow wireless LAN users to roam among service provider networks. This project was completed in late October 2002. Starting in 2003, corporate users will be able to log on at any public access hotspot identified by the new WFA logo—Wi-Fi Zone. These users will be able to gain network access via the nearest service provider regardless of where they are and yet have all charges appear on one bill from the home provider.

1.7   Accessing the Internet Without Wires

Internet access over mobile devices is becoming increasingly pervasive, particularly in consumer spaces. The Internet Protocol (IP) underpinning the Internet is a best-effort protocol in that it does not guarantee delivery of data packets. Congestion on the network can caused packets to be discarded. Confirmation of the arrival of packets at the destination is the responsibility of the Transport Control Protocol (TCP), which sits just above the IP. TCP ensures packets are received in the correct order and not lost due to congestion problems. Clearly, TCP operates well on a reliable physical link. Over an unreliable wireless channel, the errors introduced by can interfere with the congestion control mechanism of TCP.

Several methods to optimize TCP transmission over Wi-Fi wireless LANs have been proposed, including improvements in power consumption, throughput, delay, and jitter (delay variation) [3]. Some of these performance improvements require modifications only on the mobile device, thereby allowing improved performance even when a user roams into different wireless LANs.

1.8   Mobile Internet for Always-on Communication

The Internet runs on internetworking devices called routers that forward network layer Internet Protocol (IP) packets among networks with different link layers. Managing the network layer is much more flexible than the link layer since network administrators can assign structured IP addresses as opposed to unstructured MAC (link layer) addresses. However, mobile computing poses a problem if user devices with fixed IP addresses are moved from one location to another. Although a wireless LAN provides mobility support with roaming services, true mobility for large IP networks can only be realized if the IP addresses are assigned dynamically. This is because the structure of IP addressing is dependent of location and therefore needs to be changed to reflect the different locations of a mobile client. Hence, the problem of mobility in IP networks lies in the way IP addresses are structured.

The dynamic host configuration protocol (DHCP) is one method of configuring IP addresses dynamically regardless of location. A dynamically assigned IP address is known as an active lease. The active lease usually has an expiry date, which allows automatic reallocation of IP addresses that are no longer in use. Thus, DHCP relieves the administrative burden of managing IP addresses in addition to providing mobile IP addresses. When DHCP is implemented on a wired IP network, a mobile client is able to connect to the network in different locations (e.g., in different subnets). This is achieved by physically disconnecting the network cable from a fixed outlet or socket and reconnecting at the new site. Hence, ongoing connections will have to be broken when a client on a wired network moves to a new location. However, when DHCP is applied to wireless LANs, IP connections can be maintained (and applications can continue to run) even as a mobile client changes location. This removes the need to log in and out of the network. Thus, wireless LANs with mobile IP addresses can provide continuous and location-independent access to Internet services.

1.9   Your Wi-Fi Network Has No Clothes

The free-space wireless link is more susceptible to eavesdropping, fraud, and unauthorized transmission than its wired counterpart. Unauthorized people can tap the radio signal from anywhere within range. If someone sets a mobile terminal within a wireless subnet to transmit packets endlessly, all other clients are prevented from transmitting, thus bringing the network down. Being an open medium with no precise bounds makes it impractical to apply physical security like in wired networks.

Security on a Wi-Fi wireless LAN is currently built with the Wired Equivalent Privacy (WEP) protocol, which provides confidentiality and integrity of the wireless traffic. However, as it currently stands, WEP has been shown to contain flaws and does not meet the claimed security level. Moreover, the lack of a key management protocol has also brought about major concerns. For example, a 40-bit secret key is chosen manually to encrypt data and, if poorly chosen, can easily be cracked using a list of common passwords. Finally, the use of DHCP needed for mobility services does not provide strong security services.

Although the flaws in the security mechanisms used by most Wi-Fi access points are well documented, several robust solutions can be used to prevent unauthorized access to data transmitted over an 802.11 wireless LAN. Existing security mechanisms include:

- Privacy: Encrypting all data transmitted via the wireless link using longer keys. For instance, 128-bit keys are much harder to crack than current 40-bit WEP keys.
- Device Authentication: Closing the network to all clients that have not been programmed with correct network identification.
- User Authentication—important since devices can be stolen. Employing user identification/password in the network operating system.
- Implementing more powerful security measures as discussed next (e.g., network layer VPN, 802.11's TKIP and EAP solutions).

The use of secure tunnel protocols (e.g., IP Security or IPSec) over virtual private networks (VPNs) offers additional end-to-end data protection, particularly for corporate intranet connectivity over Wi-Fi wireless LANs. In this case, the end-user embeds an IPSec client and is authenticated when setting up the IPSec tunnel to the service provider (visited or home) or corporate network. Solutions are currently under development to have the end-user authenticated when connecting to the wireless LAN using DHCP and to improve DHCP security.

The 802.11i Working Group is developing a new security standard that employs stronger encryption with 128-bit keys, enhanced authentication using the Extensible Authentication Protocol (EAP), and flexible key management using the Temporal Key Integrity Protocol (TKIP). More details are presented in Chapter 3.

1.10   Simplicity Breeds Usability

Configuring a Wi-Fi network need not always be a breeze for most people, requiring fiddling and tolerance for imperfections. Some of the procedures such as dynamic addresses, network binding, and encryption can confound even the network gurus (see article by Lucky [5]). This is in contrast to the simplicity of making a call with the ordinary telephone, setting up an Internet connection with a modem or activating the cellphone by inserting a SIM card. Nevertheless, there have been increased efforts from vendors of mainstream operating systems and networking middleware in mobile and portable devices to support easy access to 802.11 wireless LANs. For example, hotspot aggregator Boingo Wireless provides a free software tool that automatically switches network profiles such as network identifiers and security keys as a user moves across Wi-Fi networks in different locations, thereby making network detection and card configuration transparent to the user.

A major reason why cellphone standards like GSM became so popular with end-users is because these standards allow cellphones to join a network without any user intervention. As public space wireless LANs become increasingly integrated with cellphone networks, end-users will demand a similar level of transparency as cellphones.

1.11   Technologies On the Horizon

As users demand higher data rates, the 802.11 committee has started focusing on the effectiveness of multiple antenna wireless systems to increase the effective capacity of existing Wi-Fi systems. Key to the potential of such systems are coding and modulation techniques, commonly called "space-time" codes, that exploit the multipath phenomenon generated by reflections of radio signals. These coding techniques rely on assumptions about the richness of the multipath distortion as well as the persistence and measurability of such effects [1].

Reconfigurable radio is a more generic wireless technology that may have an impact on wireless LANs.

Unlike the personal computer, which is a universal device that can run different applications over a common hardware, it is difficult to include new features in wireless gadgets such as cellphones and wireless LAN cards without changing the underlying hardware. Software defined radio (SDR) devices implement many functions by running software on general-purpose hardware [2], enabling such devices to switch among protocols, filtering techniques, and detection schemes. This means that at any moment, users can roam seamlessly from one wireless service provider to the next (e.g., from American to European to Japanese), keeping updated with the highest quality service, best deals, and latest radio standards, while retaining the same mobile device. Since a SDR device can be made reconfigurable by software, this makes enhanced flexibility and improved performance possible. It can also ensure more efficient use of radio spectrum and power, as well as allowing for simpler hardware design of the devices themselves.

In SDR, the analog hardware that captures and amplifies radio signals is replaced by software that implements the digital equivalents of those functions. Such an arrangement enables a single radio device to reprogram its radio mixers and filters to handle multiple modulation schemes and to work across many frequency bands. Intel and other chip makers prefer to combine all these functions into a single silicon radio chip, which costs about one-tenth as much as more advanced chips using compound semiconductors such as gallium arsenide [7]. These so-called silicon radios can incorporate multiple cellphone and wireless LAN standards, making it easier for traveling professionals to move from wireless LANs to cellphone networks without any knowledge of underlying network standards. Such radios may eventually pervade all wireless devices the same way silicon chips pervade all electronic devices in the last 3 decades.

1.12   Summary

In this introductory chapter, we provided an overview of the rapid expansion of Wi-Fi wireless LAN technology and examined both technical and business challenges for continued growth. The cost of Wi-Fi has decreased about an order of magnitude similar to the cost reduction that helped Ethernet adoption in the past. The main difference here is that being wireless, services supported by the technology are easily accessible not only to the enterprise but also to consumers and business travelers in the public and home spaces. Thus, although Wi-Fi was originally intended as an extension of enterprise Ethernet LANs, the authors envision a global landscape of affordable Wi-Fi hotspots offering ubiquitous broadband Internet access to anyone and allowing anything, from voice calls to multimedia information, to be transmitted (Figure 1.1).

In the following chapters, the authors will describe, in more detail, key technological issues related to the design and deployment of Wi-Fi wireless LANs. Armed with this technical background, we will then provide information on real statistics, case studies, a global view, and profile on suppliers, vendors, carriers, regulators, and users. This provides a basis for discussing the trends and strategies for market penetration and for evaluating the strengths and weaknesses of vendor-specific features. It is hoped that these market-oriented insights will help product managers define their product roadmaps and also define competitive hardware/software solutions for the enterprise and consumer Wi-Fi market.

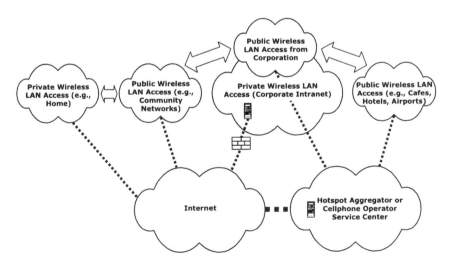

Figure 1.1:   Ubiquitous Wi-Fi Wireless LAN Access in Private and Public Spaces.

References

[1] Benny Bing, Chris Heegard, Stefan Müller-Weinfurtner, Alex Gelman, and Kwang-Cheng Chen, "Wireless LANs and Home Networks," *IEEE Journal on Selected Areas in Communications*, February 2003.

[2] Benny Bing and Nikil Jayant, "A Cellphone for All Standards," *IEEE Spectrum*, May 2002.

[3] Benny Bing, Chris Heegard, and Bob Heile, "Wireless LANs," *IEEE Wireless Communications Magazine*, December 2002.

[4] Wendy Grossman, "Wireless Wonder: A Dark-Horse Standard Could Win The Broadband Race," *Scientific American*, August 2001.

[5] Robert Lucky, "Cannot Connect," *IEEE Spectrum*, January 2002.

[6] Nicholas Negroponte, "Being Wireless," *Wired Magazine*, October 2002.

[7] Wade Roush, "Radio-Ready Chips," *MIT Technology Review*, June 2002.

[8] John Markoff, "Limits Sought on Wireless Internet Access", *The New York Times*, December 17, 2002.

[9] Eric Brown, "Wireless LANs Go Public", *MIT Technology Review*, June 18, 2001.

[10] Eric Knorr, "Mobile Web vs Reality", Special Issue on Wired and Wireless Technologies, *MIT Technology Review*, June 2001.

Chapter 2

IEEE 802.11 Standards and Applications

Unlike cellphone standards such as GSM and CDMA, the IEEE 802.11 wireless LAN standard is a network standard based on the IEEE 802 LAN architecture. This implies that an 802.11 network can be added transparently to any other 802 network, which can be based on coax, fiber or twisted pair. IEEE 802.11 standard is also popularly known as wireless Ethernet because the structure follows closely to the wired Ethernet LAN standard.

The 802.11 standard specifies the lowest two layers of a wireless LAN, namely the medium access control (MAC) and physical layers of the Open System Interconnect (OSI) network hierarchy. The standard also defines the basic architecture of such systems. It comprises a main 802.11 standard (ratified in 1997) supplemented by a suite of extensions that are indicted by the letters a, b, d, e, f, g, h, and i. These alphabets follow the time when the Task Groups for these standards are formed. The "a" group for example, was started earlier than the "b" group although the work of both groups were completed and formally approved at the same time in September 1999. Although each extension may have a different physical layer supporting a different wireless data rate, all extensions are based on a single MAC protocol. All Task Groups come under the common umbrella of the IEEE 802.11 Working Group.

There are three major 802.11 extensions operating on different frequency bands. The 802.11b and 802.11a extensions operate in the 2.4- and 5-GHz license-exempt bands respectively and are incompatible. A newer extension, 802.11g, is essentially a 2.4-GHz version of the 802.11a standard but is backward compatible with the 802.11b extension. For the beginner or the uninitiated, the IEEE 802.11 standard and its extensions can be daunting documents to read. The key to understanding the standard is to be clear about the organizational structure and the reasons behind this structure.

More information about 802.11 activities can be obtained from http://www.ieee802.org/11/. The standard can also be freely downloaded from the website.

2.1   The IEEE 802.11a Task Group

This Task Group developed the IEEE 802.11a extension, which was touted to be a natural migration from 802.11b by offering higher data rates in the cleaner 5-GHz band. It provides a maximum data rate nearly 5 times higher than 802.11b (54 Mbits/s versus 11 Mbit/s) and like its counterpart, has various fallback rates (see Section 2.2 for use of fallback rates and Figure 2.1 for the various rates). Note the 54 Mbit/s data rate is specified to be an optional rate in the 802.11a extension, which is in contrast to the mandatory 11 Mbit/s data rate defined in 802.11b. Most 802.11a vendors, however, implement the 54 Mbit/s data rate.

Among the advantages 802.11a have over current 802.11b technologies are greater scalability (because more channels are available and more fallback rates are defined for finer granularity), better interference immunity (see Section 2.2 for 802.11b interference sources), and significantly higher speed (up to 54 Mbit/s), which allows for simultaneous bandwidth-intensive applications as well as more users. The speed advantage however, has been matched by 802.11g (see Section 2.6), which unlike 802.11a, is backward-compatible to 802.11b. 802.11a also supports more channels than both 802.11b and 802.11g (12 versus 3 non-overlapping radio channels for US operation), specifically 8 indoor and 4 outdoor channels, with higher transmit powers available for the outdoor channels. This is a key advantage since more access points can be packed into the same area to support more users at higher aggregated data rates (although the need for more access points imply an increase in installation costs and complexity in network management). More importantly, the outdoor channels can be used to provide the desired range in wide-area, last-mile solutions as well as open-air networks such as community or campus networks. Recently, a new bill initiated by the Jumpstart Broadband Act calls for the FCC to allocate an additional 255 MHz of contiguous radio spectrum in the 5-GHz band (5.470-5.725 GHz) for unlicensed use. This brings the total available spectrum for 802.11a devices to 555 MHz, which is seven times larger than the 83.5 MHz available in the 2.4-GHz unlicensed band.

However, regulatory issues concerning the deployment of 802.11a networks are not completely resolved, especially their coexistence with military, aeronautical, naval, and satellite systems sharing the same 5-GHz frequency band. In addition, higher-frequency 5-GHz adaptors will require more transmitting power to achieve the same range as 2.4-GHz adaptors. The shorter wavelength of 5-GHz adaptors means that transmissions in this band have more trouble traveling through walls, floors, furniture, and other obstructions. This is mainly due to a fundamental limitation of wireless transmission—signal attenuation is proportional to the operating frequency, resulting in higher signal attenuation at higher frequencies. Thus, for a given power and data rate, the transmission range of 5-GHz 802.11a systems is limited compared to 2.4-GHz 802.11b systems. Range is a critical factor because a wireless LAN with restricted range performance implies the need for more access points to cover the same area and access points are significantly more expensive than portable wireless network adaptor cards (typically 5 to 10 times more). Clearly, a network with many access points will increase deployment costs. A solution to this constraint is to increase the operational distance of 802.11a systems at the expense of data rate performance. The question is how much data rate can be sacrificed before a breakeven point is reached where the range performance of 802.11a becomes comparable with 802.11b. This interesting trade-off issue is the subject of some recent studies by vendors (see [1]).

Another drawback of 802.11a systems is that such systems currently cannot operate in western Europe because the 5-GHz band has been reserved for another wireless LAN standard—HiperLAN2. Although HiperLAN2 employs the same physical multiplexing scheme and data rates as 802.11a, some of the parameters are different. In addition, the access methods are completely incompatible.

Unless these differences are resolved quickly, the popularity of 802.11a in Europe can be severely curtailed while the 802.11b market gains even more momentum.

2.2   The IEEE 802.11b Task Group

The 802.11b Task Group produced an extension that currently enjoys an overwhelming majority of all 802.11 wireless LAN installations. The 802.11b extension supports a maximum wireless data rate of 11 Mbit/s and is an extension of the original 1 and 2 Mbit/s 802.11 standard endorsed in 1997. 802.11b supports fallback rates of 5.5, 2, and 1 Mbit/s, which are typically used when a tradeoff in performance and range is desired. For the same transmitting power, a lower wireless data rate can increase the range due to the improved reliability in packet reception at lower rates. Note that this is in contrast to the wired Ethernet standard, which operates at a fixed data rate because the wired medium is much more reliable than its wireless counterpart. 802.11b supports a maximum of three non-overlapping channels (with a single maximum power level designed for indoor use), allowing an aggregated wireless data rate of 33 Mbit/s when three access points are co-located.

   The drawbacks of 802.11b are two-fold: it suffers from interference in the congested 2.4-GHz band and has a lower data rate than 802.11a and 802.11g. The main source of interference stems from high-power microwave ovens, which reside in practically every office and home. With new wireless personal devices involving Bluetooth starting to gain acceptance commercially, the interference in the 2.4-GHz is poised to become intolerable.

2.3   The IEEE 802.11d Task Group

This Task Group deals with regulatory issues and defines how access points exchange information on permissible radio channels and their associated power levels with end-user devices. Work has been completed and is now part of the 802.11 standard.

2.4   The IEEE 802.11e Task Group

Multimedia functionality remains an integral component of all evolving networks. Applications involving multimedia Internet access and messaging are all influenced by the emergence of multimedia. Because multimedia applications are the most general class of applications, they have the most wide-ranging traffic attributes and communication needs. In addition, these applications impose various performance requirements on the network.

There has been increased activity within the IEEE 802 standard committees to create enhancements to existing standards that will be capable of supporting multimedia communications. The IEEE 802.11 Working Group is no different. A draft standard issued by the 802.11e Task Group in November 2002 outlines how Quality of Service (QoS) and multimedia applications can be accommodated by enhancing the current 802.11 standard with new capabilities. The 802.11e extension defines eight classes of service for managing data, voice, and video applications. To accommodate this new extension, current 802.11 implementations primarily need to modify the MAC layer with no changes needed at the physical layer. The basic operation of the MAC layer is discussed in Section 2.10; more specific information on the 802.11e MAC layer can be found in Chapter 4.

2.5   The IEEE 802.11f Task Group

As mentioned at the beginning of this chapter, the IEEE 802.11 standard specifies only the lowest two layers of the OSI network hierarchy—the physical and MAC layers. However, many wireless network topologies involve packet routing among different access points, which involves higher network layers. Furthermore, the 802.11 standard does not specify the handoff mechanism to allow mobile clients to roam from one access point to another.

There have been efforts to define an Inter-Access Point Protocol (IAPP) that facilitates roaming across access points from multiple vendors under the IEEE 802.11f Task Group. The IAPP also deals with security issues and caters to wireless systems that have no wired backbone. Although standardization of IAPP has not been pursued further, the specification is available (in form of an Internet draft) and several vendors have implemented it or are in the process of doing so. Currently, no interoperability testing of IAPP has taken place.

2.6   The IEEE 802.11g Task Group

Just like 802.11a, the 802.11g extension provides data rates of up to 54 Mbit/s but operates in the 2.4-GHz band. The standard has been ratified on June 12, 2003 and is backward-compatible with 802.11b. The WFA is expected to start the interoperability and certification testing of 802.11g devices in August 2003.

2.7   The IEEE 802.11h Task Group

The 802.11h Task Group aims to enhance the current 802.11 MAC and 802.11a physical layer with spectrum and transmit power management. The major improvements include channel energy measurement and reporting, channel coverage in many regulatory domains, Dynamic Channel Selection (DCS), and Transmit Power Control (TPC). The last two features not only enable 802.11a devices to be deployed in western Europe but also play an important role in home wireless networks (see Section 4.4.2).

2.8   The IEEE 802.11i Task Group

As the name implies, the original aim of the Wired Equivalent Privacy (WEP) algorithm adopted by the 802.11 standard is to provide a level of security similar to wired networks. At the very least, WEP should be turned on, since it requires that users employ an IT-issued authentication key when accessing the wireless LAN.

The IEEE 802.11i Task Group was formed to look into improving the effectiveness of WEP and removing some security flaws. This is done primarily through MAC layer encryption and authentication over a single network link. Encryption offers additional protection as it ensures that even if transmissions are intercepted, they cannot be decoded without significant time and effort. User authentication ensures that only authorized users (over a specific device) are able to connect, send, and receive data over the wireless LAN. Data authentication ensures the integrity of data on the wireless network, guaranteeing that all traffic comes from authenticated devices only.

The 802.11i Task Group also recommends the use of a virtual private network (VPN) that increases the span of network security using the Internet Protocol (IP) network layer. VPNs provide three levels of security, namely, end-to-end (source-destination) user authentication, encryption, and data authentication. There are various forms of VPN application, including deployment in enterprise, public, and home networks.

In Chapter 3, we discuss the various security threats in more detail and describe the evolution of security enhancements to the 802.11 standard.

2.9   Physical Transmission

The various coding schemes employed by the 802.11 extensions for physical transmission are nontrivial and require some explanation. With the exception of 802.11g, the 2.4-GHz 802.11 wireless LAN extensions were all based on spread spectrum transmission. Spread spectrum refers to signaling schemes that are based on some form of coding (that is independent of the transmitted information) and that use a much wider bandwidth beyond what is required to transmit the information. The wider bandwidth means that interference typically affects only a small portion of the spread spectrum transmission. The received signal energy is therefore relatively constant over time. This in turn produces a signal that is easier to detect provided the receiver is synchronized to the parameters of the spread spectrum signal.

The Barker spreading code was employed in the first version of the 802.11 standard (issued in 1997) that operates at 1 and 2 Mbit/s. The 802.11b extension (issued in 1999) employs Complementary Code Keying (CCK), boosting data rates to 11 Mbit/s. The Packet Binary Convolutional Coding (PBCC) method is backward-compatible to both Barker and CCK codes and allows data rates to be increased to 22 Mbit/s (in 802.11b and 802.11g) and 33 Mbit/s (in 802.11g). PBCC is an option in both 802.11b and 802.11g extensions. More technical details of these coding schemes can be found in [1].

Orthogonal Frequency Division Multiplexing (OFDM) is a multiplexing technique that forms the basis of the 802.11a and 802.11g extensions. The method involves combining many radio subchannels (or subcarriers), each transporting a portion of the information contained in a data packet. OFDM is based on a mathematical concept called Fast Fourier Transform (FFT), which allows individual subchannels to maintain their orthogonality (or separation) from adjacent subchannels. This technique allows received data to be reliably extracted and multiple subchannels to overlap in the frequency domain for increased spectral efficiency. Prior to FFT, multichannel systems utilized a number of carriers generated by separate local oscillators, which proved to be inefficient and expensive.

OFDM is not a modulation scheme as some people seem to imply and does not exclude the use of other modulation schemes. In fact, 802.11a defines four different modulation schemes to be used in conjunction with OFDM. The use of OFDM is only recently becoming widespread because integrated chips that can perform high speed FFT in real time have become economical.

The primary benefit of OFDM is that it avoids multipath fading caused by signal reflections (see Section 2.11), which is a major problem in wireless communications. Each subchannel transports information at a rate slow enough so that any delayed copies due to reflections affect only a small fraction of the transmitted data. Multiple subchannels are sent at the same time, and these subchannels are combined at the receiver, thereby achieving the equivalent data rate of a single high-speed channel. When multiple antennas are employed in conjunction with OFDM, the multipath problem can be turned into an advantage, increasing available data rates significantly (see Section 2.12). The various physical transmission methods are summarized in Figure 2.1.

Data Rate (Mbit/s)	DSSS 802.11 (1997)		802.11a (1999)		802.11b (1999)		802.11g (2003)	
	Mandatory	Optional	Mandatory	Optional	Mandatory	Optional	Mandatory	Optional
1	Barker				Barker		Barker	
2	Barker				Barker		Barker	
5.5					CCK	PBCC	CCK	PBCC
6			OFDM				OFDM	CCK-OFDM
9				OFDM			CCK-OFDM	OFDM
11					CCK	PBCC	CCK	PBCC
12			OFDM				OFDM	CCK-OFDM
18				OFDM			CCK-OFDM	OFDM
22								PBCC
24			OFDM				OFDM	CCK-OFDM
33								PBCC
36				OFDM			CCK-OFDM	OFDM
48				OFDM			CCK-OFDM	OFDM
54				OFDM			CCK-OFDM	OFDM

Figure 2.1: Physical Layers of the 802.11 Standard.

2.10   Sharing Network Capacity

For wireless LANs, sharing of network capacity is essential because radio bandwidth is inherently limited. This is in contrast to wired networks, in which bandwidth can be increased arbitrarily by adding extra cables. However, the broadcast nature of the wireless link poses a difficult problem for multiple access in that the success of a transmission is no longer independent of other transmissions. To make a transmission successful, interference must be avoided or at least controlled. Otherwise, multiple transmissions may lead to collisions and corrupted signals. A MAC protocol is required to resolve these access contentions among contending clients and transform a broadcast wireless LAN into a logical point-to-point network.

The 802.11 standard currently defines a mandatory MAC protocol known as Distributed Coordination Function (DCF), which is essentially a "listen-before-transmit" protocol used in Ethernet, and an optional Point Coordination Function (PCF), which allows contention-free transmission arbitrated by the access point. In order to support multimedia applications, service differentiation and traffic classes need to be provided by the underlying wireless network. To achieve these goals, a priority-based MAC protocol is required, providing differential management for different traffic queues. The 802.11e draft extension defines a Hybrid

Coordination Function (HCF) to meet these needs. More details are provided in Chapter 4.

2.11   Multipath Fading and Delay Spread

Multipath fading is a critical problem that needs to be dealt with since it produces a variable bit-error rate that can lead to intermittent network connectivity. Multipath fading occurs even when a single client is transmitting, whereas inter-ference resulting from multiple access is due to trans-missions from two or more clients. The multipath phenomenon is caused by differently delayed versions of the original signal and this leads to a superposition of multiple signals with a delay spread at the receiver (Figure 2.2). The delayed signals are produced by reflected and scattered signals arriving at the receiver along different paths, thus resulting in differing propagation delays.

The delay spread is defined as the difference in propagation delay between the directed (line-of-sight) signal and the reflected signal that takes the longest path from transmitter to receiver. Since the delay spread is a random variable, it is often represented by its standard deviation (called the root-mean square or rms). In general, the delay spread decreases with frequency due to increased attenuation by structures in the environment. This means a 5-GHz device will encounter less delay spread that a 2.4-GHz device. Besides being affected by frequency, the delay spread typically increases with operating distance, particularly in outdoor deployments.

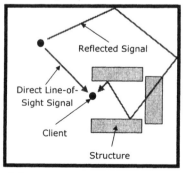

Figure 2.2:   **Multipath Phenomenon Caused by Signal Reflections.**

2.12   Next Generation Wireless LANs

The 802.11 Working Group is exploring the feasibility of providing a single, universal wireless LAN interface and extending the current data rates of wireless LANs beyond 54 Mbit/s using multiple antenna systems. As discussed in Section 2.9, current 54 Mbit/s systems, namely 802.11a and 802.11g, adopted OFDM for physical transmission. Although OFDM is inherently very robust against multipath interference, the use of antenna diversity (in other words, the use of two or more antennas placed at appropriate distances from each other) can improve the quality of the wireless connection.

The basic principle for using antenna diversity is that if one antenna receives a deeply faded signal, the other antenna(s) receive only a slightly faded version of the signal. When multiple receiving antennas are employed, detection and synchronization (both time and frequency) of OFDM signals can also be made considerably more accurate because multiple received signals from the same transmitted signal are being observed and processed. Furthermore, there is no additional transmitting power or bandwidth penalty although more extensive signal processing may lead to increased power dissipation within the device. To improve bandwidth efficiency even further, multiple transmitting antennas can be employed, giving rise to multiple input/multiple output (MIMO) systems.

These MIMO systems can be exploited for ultra-high-speed broadband wireless applications. The system capacity of such systems can be increased using different approaches: increasing the single link data rate and increasing the number of users in the entire system. A space-time processing concept can reduce the signal dimension space at the receiver by exploiting spatial correlation properties of received signals. In addition, if the signals from the antenna can be treated as independent signal components when they arrive at the receiver, then no special assumptions concerning the structure of the antenna need to be made. Furthermore, no feedback information is required at the transmitter.

2.13   Throughput versus Data Rate

Usable throughput is a key capacity concept because the overheads for high-rate wireless transmission can be tremendous. For example, the use of 1/2 rate convolutional coding in one of the mandatory OFDM modes in 802.11a and 802.11g effectively means that the usable throughput is reduced to half the raw wireless data rate. When other overheads and processing delays normally associated with 802.11 wireless transmission are included, this throughput drops further.

Many physical layer engineers tend to report capacity results in the form of *bits*/s/Hz, where bits/s refers to the raw bit rate at the physical layer and Hz refers to the unit of radio spectrum used. For high-rate wireless systems, a more accurate capacity measurement is the throughput in terms of *usable bits*/s/Hz, where usable bits/s refer to the throughput from the MAC layer and above. To put this simply, throughput takes into account all overheads associated with physical layer transmission, including all processing delays (e.g., antenna channel estimation and switching time, transmitter turn-on time, receiver signal detection time, etc.), which can be significant in high-rate wireless systems but are often ignored in data rate calculations. Note that different brands of wireless LAN adaptors can produce different throughput, even if they may all conform to the same wireless LAN standard. For example, multiple data rates are specified in the 802.11 standard, with the raw wireless rate ranging from 1 to 54 Mbit/s. The corresponding usable throughput varies (depending on the vendor) from as low as 30% to about 80% of the raw rates. This throughput decreases further and at different rates as the operating range increases.

2.14   Cable Replacement versus Mobility

First-generation wireless LANs produced in the early 1990s provided cable replacement for wireless LANs, which is not particularly suited for static desktop computers. The proliferation of notebook computers in the late 1990s changed the scenario dramatically because, unlike desktop computers, these devices, like the users themselves, are mobile. Recently, portable computing devices such as PDAs and Pocket PCs with Wi-Fi capability have also emerged. The inherent mobility advantage associated with wireless LANs provided a perfect match for personal devices. However, the mobility function requires that wireless subnets served by individual access points be properly overlapped so that continuous service can be provided as users roam from one coverage area to another.

Cable replacement has made a comeback recently, not in the office but in homes where sharing Internet connections has become a major application. User mobility in such networks is limited compared to enterprise or public access wireless LANs, hence the cable replacement model makes sense. The Wi-Fi home wireless market is driven by portable computing devices and wireless PC cards that can be used broadly in office, public, and home spaces, superceding standards like HomeRF that were developed specifically for the home environment. For instance, whereas 45% of wireless cards sold in 2000 used HomeRF, that percentage dropped to less than 10% in 2002, with over 80% of all cards being Wi-Fi products.

2.15   Wireless LAN Components

Typical wireless LAN components include wireless network interface cards, wireless access points, and remote wireless bridges.

2.15.1 Wireless Network Interface Cards

Wireless network interface cards (NICs) are not much different from the adaptor cards used for wired LANs. Like wired adaptor cards, the wireless NIC communicates with the network operating system via a dedicated software driver, thus enabling applications to utilize the wireless network for data transport. Unlike wired adaptor cards, however, these adaptors do not need a cable to connect them to the network and this allows relocation of network clients without the need to change network cabling or connections to patch panels or hubs.

Currently, PC cards developed by PCMCIA and Compact Flash cards are used to connect laptops and PDAs to wireless LANs. In some instances, these cards are embedded in the PCs and PDAs. A key limitation of these cards is that they can consume considerable power. For example, the operation of a connected PDA is currently restricted to about one hour. This can be a major impediment to "always-on" communications (in other words, seamless service across different networks) coveted by many end-users.

2.15.2 Wireless Access Points

Access points are devices that links wireless clients to the wired LAN. This enables a wireless LAN to become an extension of a wired network, inheriting all the services provided by the wired network. Because each access point creates a wireless coverage zone (a wireless subnet), wireless LANs are inherently very scalable and additional access points can be deployed throughout a building or campus to create large wireless access zones. Access points not only create wireless subnets that provide communication with the wired network, they can also filter traffic and perform standard bridging functions. The filtering function helps to conserve capacity on the wireless link by removing redundant traffic. Due to the capacity mismatch between the wireless and wired media, it is important for an access point to have adequate buffer and memory resources.

2.15.3 Wireless LAN Switches

Wireless LAN switching provides centralized control of a group of access points, simplifying management and upgrades of large-scale wireless networks. Switching technologies are a natural evolution as LAN communications achieve higher and higher speeds. By keeping all wireless traffic on different network segments, switches not only provide simpler network management but can also support bandwidth-intensive, real-time applications like voice and video well. We witnessed this evolution for wired Ethernet in the last decade when the technology moved from 10 Mbit/s to 100 Mbit/s and beyond. Now we are seeing it for wireless Ethernet as it moves from 11 Mbit/s to 54 Mbit/s and beyond. Some wireless Ethernet switches e.g., Vivato have improved the range of wireless LANs significantly and this feature alone can remove many management headaches (e.g., security, roaming, etc) in addition to reducing deployment costs considerably with the use of less access points.

2.15.4 Remote Wireless Bridges

Remote wireless bridges are similar to access points except that they are primarily used for connecting buildings in outdoor links. Depending on the distance and coverage, external antennas of differing sensitivities may be required. Such bridges are designed to link networks together, typically in different buildings and as far as 20 miles apart. They offer a quick and low-cost alternative to installing cable or leased telephone lines and are often used where traditional wired interconnections are impractical, e.g., rivers, rough terrain, private property, and highways. Unlike cable links and dedicated telephone circuits, however, wireless bridges are able to filter traffic and ensure that the connected networks are not overwhelmed with unnecessary traffic. Whereas most bridges tend to create directional point-to-point links, recently there has been interest in bridges that can deploy metro-wide wireless LAN service over hundreds of square miles with non line-of-sight (omnidirectional) coverage. Whether they are directional or omnidirectional, remote wireless bridges can present stiff competition to existing broadband local-loop solutions such as cable, xDSL, and fixed wireless. Note that although long-range 802.11 remote bridges are proliferating, 802.11's MAC protocol is optimized for shorter-range networks. This is because the carrier-sensing mechanism in the DCF protocol performs poorly with high propagation delay. This is one reason why the IEEE 802.16 Working Group, which specializes in fixed wireless for the local loop, employs a different MAC protocol from 802.11.

2.16   Wireless LAN Deployment Considerations

A wireless LAN can be deployed in an environment that is either inconvenient or impossible to install a wired network. It may offer advantages of reduced maintenance and installation costs as well as faster installation times. However, the wireless link has some unique obstacles that need to be solved. The medium is broadcast and must be shared among network clients. It can be noisy and unreliable where packet transmissions from mobile clients interfere with each other to varying degrees. The transmitted signal power dissipates rapidly in space and becomes attenuated. Physical obstructions may also block or generate multiple copies of the transmitted signal.

Wireless LAN topologies range from small, independent (or ad-hoc) networks, suitable for temporary configurations, to infrastructure networks that offer fully distributed connectivity with roaming (see Figure 2.3). Note that a wired-backbone network providing network services is a basic component of any wireless LAN deployment.

2.17   Roaming and Handoff

A key requirement for wireless LANs is the ability to handle both mobile and portable clients. A portable client is one that is moved from location to location but is only used while in a fixed location. Mobile client actually access the network while in motion. Such mobility requires a roaming function that enables a mobile client to migrate between different physical locations within the LAN environment without losing the network connection. To allow seamless roaming, adjacent wireless subnets serviced by the access points must be properly overlapped and these access points must also be able to buffer and transfer (hand off) packets to their neighboring counterparts. The complexity involved in the roaming process makes access points supporting roaming much more expensive than access points without this feature.

2.18   Health Concerns

The radiation emitted by wireless LANs is more benign than cellphones for several reasons. Unlike wide-area cellphones, short-distance wireless LANs operate at low transmitting power, typically at less than 30 milliwatts (for 802.11b). In addition, transmitting antennas are usually located at more than an arm's length away and not held close to the brain. Hence, much of the radiation will be attenuated before it reaches the user. Finally, cellphones employ continuous, connection-oriented transmissions, which are in contrast to the short, intermittent packet transmissions in wireless LANs (typically less than 1/20th of a second).

2.19   Site Survey

In a wireless LAN deployment, it is critical to determine the number of access points required to service the overall coverage area. Since access points are much more expensive than the wireless adaptor cards at the client's machine and the are more difficult to deploy than these cards due to the need for a wired connection, it is important to limit the number of access points for a given coverage area. A site survey is essential for determining the optimum number of access points.

Many wireless LAN vendors provide application programs that facilitate site surveys. Once an access point is connected, one can walk around with a wireless laptop and the programs will show signal strength and data rate at different locations and at different distances. This is helpful in determining adequate data rate and avoiding dead spots.

The accuracy of a site survey depends on several factors:

- The type of materials used in building construction and furnishings
- The density of users in a given area
- The interference level in the area
- The data rate the users need

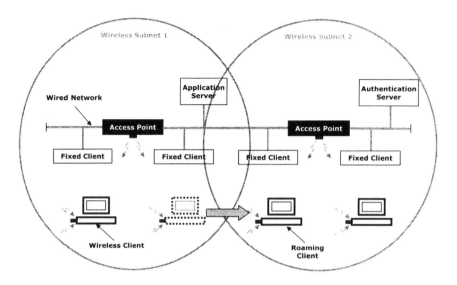

Figure 2.3: A Typical Wireless LAN Deployment.

In general, if the deployment area is large and the applications are demanding, the estimation becomes more complex. Note that using one brand of interface card and access point for site surveys can be lead to different results when a different brand is used.

2.20   Wireless Analyzers

A wireless LAN analyzer (e.g., Sniffer Wireless, WildPackets) is a passive device that listens, captures, decodes, and reports on all packetized traffic traversing the wireless network. It is a useful monitoring tool for solving network problems, including enhancing security, simplifying network management, and optimizing network performance. For example, it can detect if an intruder is repeatedly trying to authenticate with a wireless LAN. Such analyzers play an even more important role in interoperability testing and are a sound investment for start-up companies with no prior experience in developing 802.11 network adaptor cards.

2.21   Network Management

Existing management products developed for wired networks cannot be readily adapted to the management of wireless LANs. The wireless link has some unique characteristics that are very different from the wired medium. Signal quality is dependent not only on distance and transmitting power but also on reflecting objects, physical obstructions, and the amount of mutual interference from other transmitting devices. Small changes in position or direction of the antenna and shadowing caused by blocked signals and moving obstacles (e.g., people and doors) in the environment may also lead to drastic fluctuations in signal strength. Similar effects occur regardless of whether a client is stationary or mobile. In addition, external interference sources such as microwave ovens and Bluetooth devices can have a direct impact on wireless LANs. Thus, signal strength problems in wireless LANs arise from a number of factors, as opposed to component failure (e.g., broken cable, faulty network card) when such problems appear in wired LANs.

Clearly, physical layer tools are required to manage the complex signal propagation characteristics of wireless LANs. The IEEE 802.11 Working Group has chartered a Radio Resource Measurement Study Group to address this problem. Although an 802.11 device inherently avoids interference among other 802.11 devices within its vicinity by performing carrier sensing (listening to the network) before transmission, administrators can further mitigate interference among 802.11 devices by avoiding adjacent frequencies in adjacent wireless subnets while maintaining the maximum frequency spacing for each subnet. This is particularly critical in public Wi-Fi hotspot deployment where nearby establishments may not coordinate the use of 802.11 frequency channels. In this case, 802.11a has an advantage over 802.11b because 802.11a has eight non-overlapping channels for indoor use, which is more than the three channels available in 802.11b.

Unlike wired LANs, the performance of wireless LANs is much more dynamic since mobile users are more difficult to locate and track. As users move in and out of the network, and as they congregate for conferences or business meetings, the bandwidth consumption in various parts of the network can change drastically. There may be, for example, a temporary traffic overload on one portion of the network, causing its performance to suffer. Network management becomes more challenging in public hotspots where service areas can be enormous (e.g., convention centers) and bandwidth bottlenecks, security breaches, and interference from other systems can be serious problems. Thus, maximizing wireless LAN performance requires careful capacity planning followed by active monitoring and management.

An automated method of capacity management is load balancing, a feature available in most high-end access points that allows wireless LANs to serve greater loads more efficiently. Each access point can monitor the traffic load within its coverage area and then try to balance the number of clients serviced according to the traffic load in adjacent access points. To achieve this, access points must exchange traffic load information through the wired-backbone network. Most load balancing methods do not depend on signal strength as this may complicate the roaming algorithm. Typically, roaming has priority over load balancing since a mobile client must first be able to connect to an access point with reasonable signal quality and strength before load balancing can be performed.

2.22   Applications

The traditional business demand for wireless LANs lies in specific industry sectors such as healthcare, manufacturing, warehousing, education, and retail where connectivity is needed, but wired networks are impractical. Although wireless LANs also offer obvious utility in conference rooms, semi-public areas, warehouses, and other open spaces, there is disagreement about whether wireless LANs are suitable for general office environments. Deployments are discouraged because of the limited capacity (when compared to similar-cost but higher-speed wired technologies such as fast and gigabit Ethernets) and security concerns. 802.11b devices are therefore not drop-in replacements for wired infrastructure.

However, security concerns are not significantly hampering demand as enterprises can take advantage of added security features built into higher-end equipment. Since wireless LANs enable portable networks, it allows LANs to move with mobile workers that need them. By helping these workers access corporate data from home, public places or other offices, wireless LANs can lower costs and increase productivity. As an example, FedEx started using wireless LANs 5 years ago, mainly in the package sorting and aircraft maintenance areas, and the company saw a 30% jump in productivity. Interestingly, Wi-Fi technology was recently applied on Navy ships, allowing captains to command their ships anywhere onboard and eliminating hundreds of feet of cable, which add weight to the vessels.

Although the original target of wireless LANs were businesses and small office/home office (SOHO) setups looking to save money on LAN installation costs and changes, the technology has now penetrated the consumer market such as home wireless networks. Home applications require wireless residential gateways, which are similar to access points except that there are built-in modems to allow connections to the Internet. The gateway also provides wireless access for multiple users in the home, enabling shared access to the Internet, software, and printers.

Although wireless LANs are optimized for indoor environments, typically within a building, there are many instances of the technology being applied to outdoor community networks as well open-air campus networks, providing network connectivity beyond the reach of wired LANs. Just as wireless community networks are spawning hotbeds of entrepreneurial activity, many students on college campuses with wireless networks are eagerly embracing life untethered and creating an environment ideal for innovation. These networks go beyond creating an innovative environment and are subtly but profoundly altering teaching techniques, social interaction, and study habits.

2.23   Wi-Fi Deployment

2.23.1 Hotels

Luxury hotels like Sheraton, Hilton, and Marriott are offering Wi-Fi services for business travelers to enjoy high-speed Internet access in guest rooms and lobbies, or by the pool. Some 5,800 U.S. hotels are expected to offer Wi-Fi service by 2004, up from 600 today.

2.23.2 Airports

Wi-Fi networks are likely to be installed in limited areas of nearly 150 airports worldwide by 2003, approaching 300 by 2004. At least 25 of those will offer Wi-Fi service in virtually every concourse, gate, and lounge.

2.23.3 Restaurants and Coffee Shops

Cafés like Starbucks offer Wi-Fi service to customers who can surf the Internet as they sip lattes. The payoff is that customers stay longer and spend more money. Restaurants are expected to build 12,000 Wi-Fi networks by 2004, up from about 500 today.

2.23.4 Corporations

Business users in North America currently account for nearly 90% of all Wi-Fi users worldwide. Some two-thirds of the world's 1,000 biggest companies are expected to use Wi-Fi networks by 2004, up from 30% today.

2.23.5 Shopping Malls

GAP has equipped its stores nationwide with Wi-Fi so that workers can check inventory or verify prices from the showroom floor. By 2004, the largest malls in the U.S. are expected to install nearly 2,000 Wi-Fi networks so that customers can check e-mail or Web sites while they shop.

2.24   Summary

In this chapter, the activities of the various IEEE 802.11 task groups are covered. The IEEE 802.11 wireless LAN standard adopts a single MAC protocol that supports multiple physical layer extensions with multiple data rates. The standard is evolving very quickly so the reader is encouraged to consult the IEEE 802.11 Working Group website for future updates.

Wireless LANs have become an integral part of every network design, just like routers or switches. They find applications in almost any environment—industrial, government, and residential. Professionals now spend a lot of time moving around in meetings and traveling, or working from home. Wireless LANs are indispensable in allowing these mobile professionals to access computing resources away from the primary workplace, which in turn enhances productivity in a big way. There are, unfortunately, not too many scalable tools for deploying, maintaining, and managing wireless LANs in a large-enterprise networking environment. A structured approach is needed in controlling frequency channels, monitoring interference and security, checking if access points are transmitting and if clients are functioning, and taking action on all these management functions.

References

[1] Benny Bing (ed.), *Wireless Local Area Networks: The New Wireless Revolution*, Wiley, 2002.

[2] Benny Bing, *High-Speed Wireless ATM and LANs*, Artech House, 2000.

[3] IEEE 802.11 Working Group Web Site, http://www.ieee802.org/11.

[4] The Wi-Fi Alliance Web Site, http://www.wi-fi.org.

[5] Wavelink Management Tool, http://www.wavelink.com.

[6] AirMagnet Site Survey Tool, http://www.airmagnet.com.

[7] Vivato Wireless LAN Switch, www.vivato.net.

Chapter 3

Wi-Fi Network Security

Road warriors, telecommuters, and laptop-carrying executives dominate the corporate landscape, increasing demand for remote access to applications and network resources. This presents new security challenges for wireless LAN administrators as remote and mobile workers' connections create holes in network defenses. Unlike cables, radio signals are easily exposed and cannot be physically contained. In addition, the broadcast nature of wireless LANs makes it difficult to protect such LANs from unauthorized access. As such, network managers are sometimes reluctant to deploy wireless LANs unless they offer similar levels of security, manageability, and scalability as wired LANs.

3.1   Introduction

A security breach in a wired LAN is possible only if the point-to-point wired segments are physically compromised. In contrast, wireless LANs are inherently insecure because data is broadcast through the air and it is therefore hard to contain or direct to a particular recipient. As an example, "drive-by hacking" is a rising threat posed by hackers equipped with a Wi-Fi-capable laptop wandering near unprotected Wi-Fi corporate premises. Thus, although security issues can often be considered later in wired network design, it has to be considered early with wireless because of these vulnerabilities. In response, the 802.11 Working Group is aggressively developing solutions to fill the wireless LAN security hole. An interoperable 802.11 security standard that works across different clients and devices will help increase the popularity of wireless LANs significantly.

Note that securing wireless LANs is essentially the same as securing Internet connections. Although there are some wireless-specific issues related to the 802.11 standards, good basic network security practices, including intrusion prevention and detection, will overcome most concerns.

3.2   Wi-Fi Protected Access (WPA)

The evolution of network security for Wi-Fi is depicted in Table 3.1. Since the IEEE 802.11i security extension is not expected to be available until late 2003, the Wi-Fi Alliance (WFA) provided an early software fix for the major security flaws in WEP under the Wi-Fi Protected Access (WPA) initiative. WPA is:

• A strong, interoperable, security replacement for WEP

• A software-upgradeable solution to existing Wi-Fi-certified products

• Based on the portions of the 802.11i extension that can be implemented on software

Two primary security enhancements were made—an improved data encryption, which was weak in WEP, and strong user authentication, which was largely missing in WEP. It allows the WEP key to be changed faster than it can be decrypted, using the TKIP and 802.11x mechanisms, which are currently supported by the Microsoft XP operating system. When properly installed, it will provide wireless LAN users with a high level of data confidentiality and that only authorized users can access the network. Since these elements are subsets of the 802.11i draft extension, when that extension is eventually finalized, products with WPA features will be forward-compatible with 802.11i devices. At the same time, these devices will also be backward-compatible with the current installed base of Wi-Fi networks, many of which already use 802.1x authentication and TKIP encryption. A third key element of the 802.11i draft extension—the advanced encryption scheme (AES)—remains in contention within the IEEE Working Group. This prompted the WFA to move forward with TKIP and 802.1x.

In an enterprise environment, WPA can be used in conjunction with an authentication server (e.g., RADIUS) or EAP framework to provide centralized access control and management, thereby removing the need for VPNs. Because such authentication methods are typically not employed by home or Small Office/Home Office (SOHO) users, WPA can run in a Pre-Shared Key mode that allows the use of manually entered keys or passwords and is designed to be easy to set up for the home user.

The WFA have started interoperability certification testing on WPA products since February 2003. WPA has become a mandatory feature in devices bearing the Wi-Fi logo since July 2003.

3.3 The Maginot[1] Line of Wireless LAN Security (Case Study by David N. Juitt, Chief Security Architect, Bluesocket®)

The goal of this case study is to create an understanding of the issues inherent in wireless LAN security, and to propose an architecture that provides a trusted local wireless data environment.

Table 3.1: Evolution of Network Security for Wi-Fi.

	Current 802.11b Technology	Interim Upgrade on 802.11b	Complete 802.11i Solution
Authentication	Open or Shared Key	802.1x	802.1x
Key Distribution	Manual	802.1x	802.1x + Enhancement
Privacy	WEP	TKIP	AES

---

[1] After André Maginot (1877-1932), French Minister of War, who proposed a line of defense along France's border with Germany. Believed to be impregnable, the barrier proved to be of little use when theGermans attacked through Belgium in 1940.

3.3.1   The Problem

The current state of security in most wireless LANs is no more than a Maginot Line. The security mechanisms that are thought to protect these networks need to be updated to meet today's requirements. Undue trust has been placed in the present industry standard mechanisms that purportedly protect our proprietary resources from access via wireless data methods.

In February 2001, the Wireless Ethernet Compatibility Alliance (WECA) issued a statement [4] responding to the problems in the state of wireless LAN security. The statement was made in response to the first of many attacks that would succeed against Wired Equivalent Privacy (WEP), the privacy and authentication mechanism used in 802.11 wireless LANs.

A team of researchers at University of California at Berkeley correctly reported on "a *sophisticated* methodology [that] could be potentially used to compromise security of the Wired Equivalent Privacy (WEP) mechanism of Wi-Fi (IEEE 802.11b standard) wireless LAN products" ("Intercepting Mobile Communications: The Insecurity of 802.11" [5]).

In the months to follow, several attacks and analyses were performed that seriously stalled the advancement of an interoperable security solution for Wireless LANs. A series of papers from Intel, the Department of Computer Science at the University of Maryland, and the Computer Science Department of the Weizmann Institute in conjunction with Cisco have pointed out numerous flaws in the WEP standard. The Cisco/Weizmann paper delivers the most devastating blow to WEP by describing a passive attack (i.e., one in which the attacker merely listens to data passing over the network) that can be used to recover a WEP key. Adam Stubblefield, John Ioannidis, and Aviel D. Rubin then demonstrated that 128-bit keys could be recovered trivially using the Fluhrer, Mantin, and Shamir (FMS) attack. This case study looks at the current landscape, how we got here, and how the technology needs to improve.

3.3.2   Security Issues Affecting Wireless LANs

As in any environment in which data is transmitted over untrusted media, certain safeguards must be in place and effectively managed. In the case of wireless LANs, the fundamental trust model is different due to the inability to rely on basic properties of the media to provide rudimentary protection mechanisms. More simply put, wireless LANs promote casual eavesdropping purely because they don't use a wire.

In order for a wireless LAN security system to be effective, it needs to address the following three security issues:

- Providing access control: authenticating users and authorizing them to access particular resources, while denying access to unauthorized users

- Ensuring link privacy and integrity: preventing unauthorized users from reading, introducing, or altering data transmitted over the network

- Preventing denial of service: ensuring that one user, or a small group of users, cannot take up all of the bandwidth available on an access point and so deny access to other legitimate users

The following sections describe how the 802.11 standard performs access control and attempts to ensure privacy. Denial of service is not addressed at all in the current 802.11 standard.

3.4   Initial 802.11 Security Approaches

3.4.1   Authentication

Each node in a standard 802.11 network uses a network name (referred to as the Service Set IDentifier or SSID). Before associating with a particular access point (AP), users can be required to enter the AP's SSID together with a password. Unfortunately, the SSID is regularly broadcast by the AP and can easily be detected. The password is sent unencrypted and unlimited retries are allowed. This weakens the use of an SSID as an authentication token.

The Medium Access Control (MAC) address of the Wi-Fi card can also be used for authentication and authorization. The MAC address is a unique 48-bit serial number allocated to each Ethernet type device. Although it is not part of the 802.11 standard, many vendors allow the use of access control lists. Lists of authorized MAC addresses can be stored inside each AP. Only client devices whose MAC addresses appear in an AP's access control list are allowed to connect to that AP. The problem with this approach is that an attacker can easily "sniff" the MAC addresses of clients that are connected to an AP, since they are never encrypted. Most Wi-Fi client, devices allow the MAC address to be changed via software, so once in possession of a valid MAC address, the attacker can easily masquerade as a legitimate user.

3.4.2   Wired Equivalent Privacy (WEP)

In addition to the measures described above, the 802.11 standard defines a combined access control, data privacy and data integrity system called Wired Equivalent Privacy (WEP). These aspects of WEP are described in turn below.

3.4.2.1 Access Control

WEP can be configured on a Wi-Fi network so that a user cannot gain access without the correct key. This symmetric encryption key ranges from 40 to 128 bits. Generally, the key is physically typed into each device and then stored for future use. Since the same key is used on each mobile device, if one key is compromised (e.g., a notebook computer is stolen) then all the remaining devices need to have their keys changed. WEP does not currently provide this key management function.

3.4.2.2 Link Privacy

The 802.11 standard outlines the use of WEP as the encryption algorithm to ensure privacy in a wireless LAN. The WEP encryption algorithm is based on a standard known as RC4 that was developed in 1987 by Ron Rivest for RSA Data Security, Inc. The IEEE 802.11 group chose RC4 for WEP because it was cheap to license and easy to implement in software or hardware, thus allowing the vendors who made up the group to bring products to market quickly at affordable prices. Moreover, RC4 was freely exportable from the U.S., providing the key length was limited to 40 bits. RC4 is regarded as a reasonable encryption system for its time, but it can no longer be described as state of the art. RC4 is a rudimentary stream cipher, and must be implemented in an appropriate manner.

3.4.2.3   Link and Data Integrity

The standard way of ensuring integrity is to append some form of message authentication code to each block of data before it is transmitted. In WEP, the transmitting device generates a 32-bit cyclic redundancy code (CRC-32) checksum by performing a polynomial calculation on each frame of data being sent. The checksum is appended to the data frame. The receiving device performs the same polynomial calculation on the data and if the answer matches the checksum it has received, the data is assumed to be uncorrupted. Again, this is a fairly basic approach but one that is simple to implement.

3.4.3   WEP's Fatal Flaws

The reason WEP failed was the attempt to use the RC4 stream cipher for both the authentication and privacy functions. Even though RC4 is a perfectly good encryption algorithm, it was not applied correctly. RC4 explicitly warns never to use the same key material twice, no matter what the payload, since it is simply an XOR stream cipher.

The fatal flaw is the 24-bit long initialization vector (IV) that gets transmitted in the clear in every packet. Two straightforward changes (using a longer IV and using a secure hash algorithm instead of CRC-32 for integrity) might have avoided the weakness in the privacy portion of the implementation.

The generation of the key schedule for RC4 involves appending the IV to the WEP shared key. In high-traffic wireless environments, many packets are dropped, requiring a resend. Every resend of a packet is supposed to change the IV (which only has a 2^{24} keyspace), but often that is not the case. Walker's study [4] claims that two packets will have a 99% probability of a key space collision after only 2^{12} frames. This claim translates to < 5 seconds in a loaded 11 Mbit/s network; every few seconds there will be more than one packet transmitted using the same (IV, WEP) key pair. Obviously, the problem will be exacerbated in 802.11a when bandwidth rises to a 54 Mbit/s maximum.

The attack involves collecting lots of traffic, the associated plaintext fields (IP addresses and sequence numbers for instance), and IVs. Every time there is an IV reuse, more information is available to do the cryptanalysis. Since we know some of the information in the packet (the plaintext), we get to generate a table that maps out the XORing material. This is a basically a master decryption table. Demonstrations show that this table can be created in a matter of hours. This is not as sophisticated an attack as WECA would like wireless LAN users to believe.

Appending the IV to the WEP key was a flawed design. 40-bit WEP keys are provide very little protection. 128-bit keys will just require more traffic to be collected to perform the attack.

3.4.4   802.1x

The IEEE has always been aware of the security implications and implementation headaches of the WEP symmetric key approach. Cisco and Microsoft proposed a solution called 802.1x that provides access control on any Ethernet port–wired or unwired. It is currently awaiting ratification by the IEEE standards body.

The 802.1x solution still uses the WEP algorithm but now generates and distributes a new symmetric key each time a mobile device connects to an AP, so there is a new key per user session. This dramatically reduces the amount of data transmitted with the same key, and so largely defeats the type of attack employed by the University of Berkeley group.

802.1x uses a RADIUS/LDAP server to authenticate the users and manage the keys. The authentication data (user name and password) are transported using LEAP (Lightweight Extensible Authentication Protocol). The new WEP key is transmitted to the mobile device by encrypting it with the WEP key used for the last session or by a default WEP key.

3.5   Is the Problem Intractable?

3.5.1   Wireless Networks

One reason for a different solution is the realization that wireless LANs are completely different environments from wired LANs. The initial approach for providing security in wireless LANs was to create the equivalent of wired privacy. That is a necessary element, but not a sufficient solution.

Link layer security is designed to provide a protection mechanism from one point to another. The base assumption is that any data that travels on that link is considered to have been "created equal," i.e., there is no inherent difference in any of the transmitted data. When used in a wireless LAN, link layer (Level 2) security only provides the ability to make coarse-grained (on or off) decisions regarding access to the network.

A wireless LAN is much more so a traditional client/server environment than a link extension of a wired LAN. Wireless networks require the ability to identify accurately the source of the data and make a decision as to whether that traffic belongs on that link. In essence, none of the data can be treated from a bulk perspective.

3.5.2   The Need for a Unified Approach

Traditional infrastructure solutions for providing authentication, authorization, service control, and security management in a wireless LAN have required the configuration and integration of several different network components. With each of these components, there is often a different user interface. This can lead to a high degree of complexity and risk of misconfiguration when deploying a wireless data solution.

Similarly, end-users may or may not care about security. In many cases, their employer cares a lot more than they do. If the security system is too hard or cumbersome to use, then they will seek to circumvent it. Making the security as transparent as possible is therefore a key design goal.

3.5.3   The Need for Key Management

One of the fatal flaws in WEP can be traced to the lack of a key management mechanism. Early work by the IEEE 802.11i Security Task Group had proposed Kerberos as a solution.

Kerberos has it's own problem. It is an extremely complex solution that depends on trusted (physically secure) online servers and time stamps for replay detection (requires time synchronization between all APs and mobiles and a physically secure clock). Hence it does not lend itself well to being implemented in $250 APs left in unprotected locations.

3.6   A Comprehensive Security Architecture for Wireless LANs

Wireless LANs need to provide the following security controls.

3.6.1   Providing Improved Access Control

Authentication of end-users is a technique that has matured over the evolution of remote access systems. Authentication is the ability to identify the user of the service requested. Using state-of-the art technologies, several mechanisms can be deployed in the same system. Authentication needs to be performed at the user level, not the machine level. Ideally, the security architecture will support multifactor authentication that corresponds to the value of the information that needs to be protected. Strength of authentication is tied to the use of multiple factors of information, i.e., something you know (a password), something you have (a hardware token card, a digital certificate), or something you are (biometrics).

Authorization is the process of determining whether an authenticated user has the appropriate privilege to access a requested resource. The union of authentication and authorization is commonly referred to as access control.

One of the most challenging problems in managing large networked systems is the complexity of security administration. Today, security administration is costly and prone to error because administrators usually specify access control lists for each user on the system individually. Role-based access control (RBAC) is a technology that is attracting increasing attention, particularly for commercial applications, because of its potential for reducing the complexity and cost of security administration in large networked applications.

With RBAC, security is managed at a level that corresponds closely to the organization's structure. Each user is assigned one or more roles, and each role is assigned one or more privileges that are permitted to users in that role. Security administration with RBAC consists of determining the operations that must be executed by persons in particular jobs, and assigning employees to the proper roles. Complexities introduced by mutually exclusive roles or role hierarchies are handled by the RBAC software, making security administration easier.

3.6.2 Ensuring Link Privacy and Integrity

3.6.2.1 Tunneling

Protocols such as IPSec and PPTP need to be used to provide the richness of key management and encryption algorithm negotiations. Native compression capabilities in these protocols can also provide increased performance in the wireless environment.

Table 3.2: Throughput Measurements in Wireless Networks.

Privacy mechanism	Data transferred	Bandwidth
64-bit WEP encryption	4.6 MB	3.7 MB
None	5.8 MB	4.7 MB
IPSec	7.0 MB	5.6 MB

3.6.2.2 Detecting Fake Access Points

Preventing intruders from connecting their own APs to the network with the intention of capturing users' passwords and mounting a "man in the middle attack."

3.6.2.3 Detecting Rogue Access Points

Preventing users from installing unauthorized access points, which, through poor configuration or design, could act as "back doors," providing unprotected access to the corporate network.

3.6.2.4 Optimized for Wireless Links

Portability is another fundamental difference between wired and wireless networks. In order to make the security solution transparent and meet user expectations, it must provide seamless authenticated mobile handoffs anywhere in the enterprise.

3.6.2.5 Preventing Denial of Service

Due to the potential bandwidth bottlenecks introduced by wireless network architectures, it is of paramount importance to ensure that one user, or a small group of users, cannot take up all of the bandwidth available on an access point and so deny access and acceptable network performance levels to other legitimate users.

3.7   Summary

The biggest advantage for wireless LANs–open access for people that come and go–is also the biggest security disadvantage. Despite its flaws, WEP provides some margin of security compared with no security at all and remains useful for the casual home user for purposes of deflecting would-be eavesdroppers. For large-enterprise users, WEP native security can be strengthened by deploying it in conjunction with other security mechanisms such as VPNs or 802.1x authentication with dynamic WEP keys.

In this chapter, the essential elements of designing an effective security system for wireless LANs have been identified. These include:

- Minimizing the security threats of lost or stolen hardware, rogue access points, and hacker attacks

- Employing user-specific, session-based WEP keys created dynamically at user logon, not static WEP keys stored on client devices and access points

- Managing the security for all wireless users from a central point of control

VPN is an established security overlay that can cure the current problems of Wi-Fi without relying on 802.11i activities but the implementation complexities, particularly on a corporate premise, may prove unacceptable. WPA is an early, prerelease version of the IEEE's 802.11i standard meant to replace WEP. WPA's interoperable security enhancements increase the level of data protection and access control for existing and future wireless LAN systems. Designed to run on existing hardware as a software upgrade, WPA is derived from and will be forward-compatible with the upcoming IEEE 802.11i standard. In essence, a wireless LAN can be treated like an Internet or dial-up connection with the use firewalls, VPNs, authentication, and encryption to secure wireless access.

References

[1] "IEEE Standard for Local and Metropolitan Area Networks—Port-Based Network Access Control," IEEE Std 802.1x-2001, June 2001.

[2] National Institutes of Standards and Technology white paper on wireless security best practices— http://csrc.nist.gov/wireless/S05_NIST-tk2.pdf.

[3] Wi-Fi Alliance Web Site, www.wi-fi.org.

[4] Wireless Ethernet Compatibility Alliance, *802.11b Wired Equivalent Privacy (WEP) Security,* February 19, 2001.

[5] N. Borisov, I. Goldberg, and D. Wagner, *Intercepting Mobile Communications: The Insecurity of 802.11*.

[6] J. Walker, *Unsafe at any key size; An Analysis of the WEP Encapsulation*.

[7] http://csrc.nist.gov/rbac.

Chapter 4

QoS Provisioning for 802.11 Wireless Home Networks

New electronic gadgets and services for home applications are proliferating, including multimedia digital set-top boxes, personal video recorders, mobile Web pads, residential gateways, wireless MP3 players, digital audio/video jukeboxes, video-on-demand, unified messaging, high-speed Internet sharing, and Internet telephony. Recently, Apple Computer unveiled the new iMAC that has a compact computer at the base that could act as a home entertainment server. At the same time, Microsoft introduced the Mira flat control panel that comes with a Wi-Fi wireless interface and can be used to regulate entertainment and work. These devices are increasingly being networked to enable communication and to allow broadband services to be delivered to the home, creating Internet appliances and forming digital hubs that unite home computers, consumer electronics, and audio/video systems [10].

A key problem for deploying 802.11 wireless home networks is in servicing the wide array of entertainment traffic. The major traffic types can be summarized as follows:

- Best effort (e.g., email, Web browsing, file transfer), in which no guarantees are required with respect to bandwidth, latency, or even eventual delivery
- Audio streaming (e.g., MP3), which is medium rate and exhibits very low tolerance to reception errors
- Video streaming (e.g., MPEG), which is high rate and imposes limits on error rates, maximum delay, and delay variation (jitter)
- Interactive gaming, which is typically low rate and delay-sensitive

- Voice, Internet telephony, and video conferencing, which are low to medium rate but very sensitive to delay and delay jitter
- IEEE 1394 FireWire applications (e.g., digital TV), which are very high rate (hundreds of Mbit/s) and very sensitive to uninterrupted isochronous (extremely low delay jitter) operation

Clearly, the applications that support these traffic types impose diverse performance requirements on the network. Real-time applications involving voice, video, and audio are particularly demanding because these applications have strict timing bounds. A maximum delay is allowable after which the information generated is no longer useful and can be discarded. On the other hand, early packet arrivals can lead to buffer overflow because the previous information has not been played out and are still in the buffer space.

4.1   Basics of Quality of Service (QoS) Provisioning

For advanced or time-critical applications, QoS provisioning can fine-tune traffic flow, ensuring efficient bandwidth management when delivering voice and video traffic and satisfactory user experience. There are three general methods of providing QoS. The most straightforward method is to add more bandwidth (capacity) to increase the usable throughput (or effective information rate) but this may require a network upgrade, which is not always possible (e.g., wireless radio spectrum is inherently finite and therefore limited). Furthermore, upgrading the bandwidth alone is not sufficient since subscribers buy services and the contents that come with them, not just bit rates.

The next method employs data compression and coding that reduce bandwidth requirements and the likelihood of packet loss, at the expense of increasing overall delay and possibly delay variation. However, this process is not necessarily the opposite of bandwidth expansion and may not be suitable for live streaming applications or the transmission of sensitive documents such as medical X-rays and CT scans.

The final method involves defining aggregated traffic-dependent performance metrics that guarantee resources for key applications in a timely manner. These QoS metrics can be expressed in terms of allowable data throughput, maximum delay, delay variation (jitter), packet loss rate due to buffer overflow (typically less than 1%), packet error rate due to corrupted transmissions, and service availability (typically 99.9999%).

Just as applications in a computer will run faster when the computer memory resources are utilized optimally, well-specified QoS parameters help network planning through measurement and management of traffic flows (or traffic streams), enabling more efficient bandwidth allocation and utilization, which can be critical in bandwidth-limited wireless networks or low-speed, wide-area networks (WANs). For these reasons, QoS has become important, and the contracts that specify it (called service level agreements or SLAs), are becoming very common. The usual agreement specifies the end-to-end performance to which the client is entitled over a specified period of time. Defining QoS parameters is only a first step. There must also be mechanisms to label traffic flows with respect to their priorities (to distinguish traffic characteristics), and for the network to recognize and act on those labels (in order to service connections with different QoS).

4.2   QoS Provisioning in Home Wireless Networks

The major QoS issues associated with home wireless networks are summarized here. Essentially, a dynamic resource allocation technique is needed, using a combination of admission control, traffic shaping, and policing mechanisms to assign bandwidth, channel, delay, and power levels based on current interference, propagation, and traffic conditions.

4.2.1 Reserved Bandwidth

Bandwidth must be guaranteed for an application not only to satisfy the bandwidth requirement but also to limit the transmission errors and guarantee the delivery of data packets at some specified maximum latency, with some amount of tolerance in the spread of latencies. Note that when there is excess bandwidth, both best effort and reserved bandwidth service work well. However, when there is network congestion, best effort service slows down all users while reserved bandwidth service block some users in order to fully serve others. For applications involving real-time voice and video, a reserved bandwidth is preferable in many instances [1].

4.2.2 Error Control

If a link is unreliable (e.g., wireless), then some of the QoS issues need to be addressed at the link layer (in the local network) rather than from an end-to-end basis (involving remote networks). A combination of forward error correction (FEC) coding and automatic repeat request (ARQ) is effective in improving QoS.

FEC techniques typically use error-correcting codes (e.g., convolutional coding) that can detect with high probability the error location. These channel codes improve the bit error rate performance by adding redundant bits in the transmitted bit stream that are employed by the receiver to correct errors introduced by the channel. Such an approach reduces the signal transmitting power for a given bit error rate at the expense of additional overhead and reduced data throughput (even when there are no errors). Employing FEC alone is not effective since packets may be lost or dropped due to buffer overflow at the receiver.

In ARQ, the receiver employs error detection codes (e.g., cyclic redundancy check) to detect errors in the received packets and then request the transmitter to resend any corrupted packet. It is simple to implement and is most useful when the link characteristics are unknown or unpredictable. For channels with feedback, the ARQ scheme perfectly adapts to channel conditions. However, the ARQ technique may degrade packet delay variation performance. Interactive applications also impose a limit to the number of ARQ retransmissions.

In general, applications that are not delay-sensitive can be transmitted with ARQ, whereas real-time applications involving voice and video are better transmitted with FEC.

4.2.3 Resource Allocation

If bandwidth is constrained (e.g., wireless), it may be necessary to reject the admittance of additional traffic to preserve outstanding connection agreements. The admittance decision is usually based on several criteria such as:

- Traffic characteristics
- Connection holding time statistics
- Desired QoS of each traffic class
- Amount of resource available

A typical strategy is to allocate resources for real-time applications first before dedicating residual resources for best-effort services. In addition, if absolute guarantees are not possible, then it may be sufficient to provide predictive or approximate service, indicating that the application's requirements are most likely to be met.

4.2.4 Traffic Shaping

The communication medium may be encoded by algorithms that generate different data rates. Efficient video coding schemes (e.g., MPEG) that can maintain the same quality for all images normally produce variable bit rate (VBR) outputs. Figure 4.1 shows how a bursty VBR video traffic source can be shaped (adapted) to a constant bit rate (CBR) traffic stream using a smoothing buffer. Since each video frame is compressed with a different ratio, the frame boundaries after shaping become irregular. However, because the average bit rate of the shaped VBR video stream is now lower than the peak rate of the uncompressed CBR video stream, the number of simultaneous users is increased.

Generally, a wireless link is most efficiently shared among devices with CBR requirements. It is also important to realize that voice and video communications are useful at various levels of quality. For example, corrupted packets are more noticeable in an audio stream than in a video stream since people can communicate solely using voice but the video component alone may not be satisfactory. Similarly, real-time streaming applications involving audio or video streams are more tolerant of errors than data applications. These observations imply that it is more important to increase the probability that most packets arrive by their expected deadlines than to achieve the successful arrival of all packets with no regard to delay.

4.2.5 Adaptive Applications

Besides unequal levels of error protection, applications must also have the ability to adapt quickly to changes in the available bandwidth or network delay. If applications running on different user devices are designed to auto-matically compensate for link impairments independently and without intervention from the source, graceful service degradation is ensured as the quality of the wireless link deteriorates. Such a function is particularly useful for the multicast of multimedia information, in which link errors and transmitting power considerations limit the effectiveness of sustained wireless broadcasting. Essential requirements to support adaptive applications include:

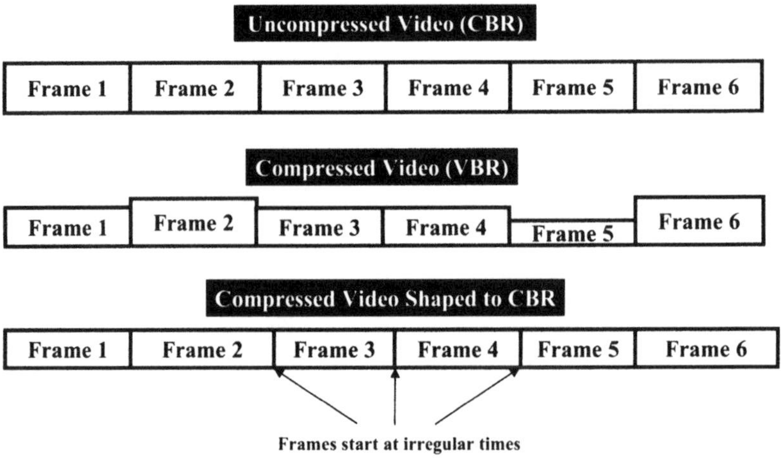

Figure 4.1: Traffic Shaping.

- Memory buffers at the receiver
- Flow control if the receiver does not have sufficient buffer capacity to accommodate all data received
- Processors to handle intersample spacing and playback point at the receiver
- Layered (hierarchical) coding e.g., multiresolution (scaleable) source coding
- Error concealment techniques based on spatial or temporal interpolation from the adjacent areas of the same frame or the previous frame (these techniques require detection of packet loss in order to locate the damaged areas of the image)
- Cross-layer design for channel and QoS adaptivity, in which strict layering rules of the traditional layered architecture are violated in order to share more information across different protocol layers [5]
- Traffic prediction and control

4.2.6 Media Compression

Media compression schemes rely on two basic principles—the removal of redundancy and the reduction of irrelevancy in the input data. Although lossless compression, in which the original signal can be completely restored upon decompression, is used in some archival, legal, and medical applications, the pragmatic goal in most media applications is some degree of perceptual losslessness [7].

Recent research has investigated the adaptation of compression algorithms to changing link quality. The motivation for such joint source/channel coding came about due to the unpredictable and imprecise information provided by the wireless link. The overall capacity is divided dynamically between the compression algorithm and channel coding. Optimal performance can be achieved in the following manner. If the quality of the link is good, all the capacity is allocated to the compression scheme and no channel coding is needed. As the link quality degrades, increasing levels of capacity are allocated to channel coding to deal with errors caused by the link. Such a cross-layer design approach may have implementation difficulties because source and channel coding techniques are contrasting technologies—source coding removes redundancy, whereas channel coding typically adds redundancy. Hence, they are traditionally developed separately. This separation principle dates all the way back to Claude Shannon's seminal work on information theory in 1948; it allows source compression to be developed separately from channel coding as long as the source encoder produces a bit rate that can be handled by the channel encoder.

4.2.7 Impact of Higher Layers

As described briefly in Section 1.7 of Chapter 1, confirmation of the arrival of data packets at the destination is the responsibility of the Transport Control Protocol (TCP), which sits just above the Internet Protocol (IP). If any packet is not delivered (as determined by tracking the sequence numbers of packets at the destination), TCP requests a retransmission of the missing packet, thereby ensuring that all packets eventually get to the destination. This is effective, but slow. Therefore, TCP is generally used by applications that are not time-sensitive. Internet telephony, for example, will not work properly. Such applications rely on what is essentially a stripped-down version of TCP, known as the user datagram protocol (UDP), which runs faster than TCP by omitting some of its functionality. Applications that run over UDP must either have those missing capabilities built into them or do without them [3].

Besides not being able to service real-time traffic properly, the performance of conventional TCP over an unreliable wireless link is also poor because TCP's congestion avoidance algorithm (which is controlled by the sender) works best only for networks that experience low likelihood of packet loss. TCP makes the implicit assumption that retransmission is a result of network congestion. Thus, retransmissions at the wireless link layer can be perceived as periods of heavy losses by the transport layer, causing retransmission timeouts. In such cases, TCP reacts by drastically reducing the current transmission rate. It first reduces the transmission window size to restrict the amount of data flowing through the network and then activates the slow-start mechanism that lowers the incremental rate of the window size to only one packet. Finally, TCP resets the retransmission timer to a backoff interval that doubles with each consecutive timeout. These measures reduce the load on intermediate links, thereby controlling congestion on the network. However, TCP takes a long time to recover from a transmission rate reduction that results in severe throughput degradation.

Such problems can be mitigated through a TCP-aware link layer in which the residential gateway triggers local retransmission. This approach attempts to make the lossy wireless link appear as a higher-quality link with a reduced effective bandwidth. As a result, most of the losses seen by the TCP sender are caused by congestion. Another method attempts to make the sender aware of the existence of wireless links and realize that some packet losses are not due to congestion [1]. The sender can then avoid invoking congestion control on non congestion-related losses.

4.2.8 Voice Traffic Support

Packetized voice transmission over the Internet has revolutionized the way telephony traffic is carried over networks. The technology is driven by a tremendous reduction in deployment costs regardless of distance. The service is popular with consumers because it offers a high-quality voice connection for a fraction of the cost of traditional lines on a public switched telephone network. The worldwide market for enterprise IP telephony, including IP phones, hit $171 million in the second quarter of 2002.

A distinction needs to be made between voice over the Internet and voice on a private IP network. Although one can build a private network using QoS over Ethernet and QoS-aware routers, voice on IP (VOIP) will generally work fine. However, voice over the Internet (VoIP) is quite different simply because there are no guarantees that a voice packet will be prioritized and managed better over a public network.

4.3   QoS Support at the Higher Network Layers

The Internet Engineering Task Force (IETF) proposed several methods for improving QoS, including Integrated

Service (IntServ) and Differentiated Service (DiffServ). IntServ is the earlier method and allows the network to support multimedia traffic using IETF's Service Level Agreements (SLAs). These SLAs provide explicit reservation and parameterization of traffic on a per-flow basis, effectively assigning a specific flow of data to a traffic class, which defines a certain level of service. Another form of IntServ supports traffic parameterization on a per-packet (rather than per-flow) basis. This form allows greater control over the differentiation of traffic types than that provided by simple prioritization while avoiding the complexity of flow parameterization.

A reservation protocol (most predominantly resource reservation protocol or RSVP) is used to communicate information about a flow to the admission controller, and then to the packet scheduler after connection is established. RSVP operates on the premise that options for QoS are needed but not as a replacement for best-effort service since many applications cannot predict in advance what their bandwidth requirements will be. Once the data flow is assigned a class, a path message is forwarded to the destination to determine whether the network has sufficient resources to support that specific class of service. If *all* devices along the path are found capable of providing the required resources, the receiver generates a reservation message and returns it to the source. The procedure is repeated continually to verify that the necessary resources remain available. If the required resources are not available, the receiver sends an RSVP error message to the source.

IntServ has several problems. There is no guarantee that the necessary resources will be available when desired. It also does not scale well since it reserves network resources on a per-flow basis. If multiple flows from an aggregation point all require the same resources, the flows will all be treated independently and the reservation message must be sent separately for each flow [3].

In DiffServ, any traffic management or bandwidth control mechanism that treats different flows differently (e.g., weighted fair queueing, RSVP, any lightweight mechanism that does not depend entirely on per-flow resource reservation) can be employed. A short tag is appended to each packet depending on its service class.

Traffic flows having the same resource requirements may then be aggregated on the basis of their tags when they arrive at the edge routers. The routers at the core then use the tag information to forward the flows to their destinations. This is done without examining the individual packet headers in detail, allowing the core network to run much faster. The present trend in providing QoS for IP networks is to use DiffServ complemented by some of the resource reservation capabilities of RSVP.

Diffserv and Intserv need support from layers above the transport layer in order to specify or maintain QoS. The upper layers either tag the traffic appropriately or create a path for the data flow and specify path parameters. Thus, the method of identifying and classifying users, applications, and network resources is important. A policy-based approach employs the common open policy service (COPS) and lightweight directory access protocol (LDAP) standards. A client can add advanced services easily using COPS, which specifies a service in unequivocal terms and allocates the resources required to deliver that service. The tool is more adaptable to a customer's requirements, allowing those requirements to vary with time of the day, application, or even user session. The requirements and rules for resource allocation (known as policies) are decided in advance. Once a user has put such a policy in place, network parameters can be configured to meet customer-initiated QoS requirements.

The ability to measure and display QoS parameters is important in assuring customers the service they are paying for. TCP/IP protocols are well suited to measurement of metrics like throughput, forwarding rate, and packet loss. Such feedback to users on actual network usage, even in the absence of usage billing, can encourage efficient and prudent use of network resources.

4.4   QoS Support in IEEE 802.11 Wireless LANs

Although there are competing home wireless standards such as Bluetooth and HomeRF, the IEEE 802.11 standard is by far the most widely adopted. However, the current standard cannot support QoS in an adequate manner. For example, the Distributed Coordination Function (DCF) based on Carrier Sense Multiple Access with Collision Avoidance (CSMA/CA) requires mandatory transmission of acknowledgments (a form of ARQ) for every packet correctly received (Figure 4.2). Furthermore, senders activate the backoff mechanism even when there are no collisions, a mechanism designed to avoid collisions immediately after a successful packet transmission (Figure 4.3). The ARQ and backoff mechanisms are not necessary when servicing real-time traffic and, depending on the number of contending users, may degrade performance due to decreased throughput and increased delay and delay jitter.

In this section, we address some of these issues by focusing on the proposals in task groups e, h, and i within the IEEE 802.11 working group, which have a direct impact on QoS provisioning for home wireless networks.

4.4.1 IEEE 802.11e

This task group aims to enhance the existing MAC functions by adding multimedia capabilities, QoS support, and home wireless network features. The proposed changes will operate with the existing 802.11b as well as new high-rate extensions such as 802.11a and 802.11g. The new proposals include [2]:

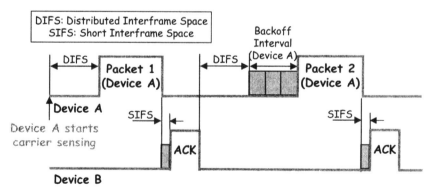

Figure 4.2:   Mandatory Acknowledgments and Back-off using 802.11's DCF MAC Protocol.

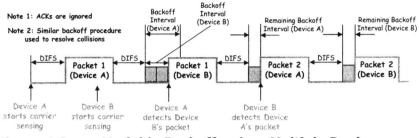

Figure 4.3:   Variable Backoffs when Multiple Devices Transmit using 802.11's DCF MAC Protocol.

- Allowing a device to communicate directly with another even in the presence of an access point
- Removing and aggregating acknowledgments for packets in a traffic flow
- Using Reed–Solomon coding for forward error correction;
- Allowing a device to transmit multiple packets within a specified period
- Contention-free polling during the contention period
- Efficient polling using reservation requests
- Priority-based contention backoff
- Traffic prioritization and parameterization (scheduling algorithm likely to remain unspecified)

The main motivation for these changes is to reduce the overall (end-to-end) delay and delay variability (jitter) for real-time traffic such as voice and video. For example, any transmitted voice packets that have been corrupted by the wireless channel can be dropped (and not retransmitted using ARQ) to reduce the overall delay. The use of individual ACKs increases the delay and can therefore be removed or aggregated when service real-time traffic. As a rule of thumb, voice and video transmission can typically tolerate an end-to end delay of up to 100 ms.

It is important to emphasize that these changes need not necessarily mean that the current 802.11 standard (without the 802.11e extension) is unable to service real-time traffic. If the number of users serviced is low (which is typical of a home network), several real-time traffic streams can be serviced. A recent research study has shown that, theoretically, it is possible to service up to 30 voice traffic streams using 802.11's native DCF MAC protocol at 1 Mbit/s.

Note that the data rate also plays a crucial role here. A high data rate (e.g., 54 Mbit/s) can reduce the delay significantly since the transmission time for the real-time packets becomes considerably shortened.

4.4.2 IEEE 802.11h

Spectrum management is an important aspect of QoS support since wireless error rates can vary dramatically when obstacles (e.g., doors, people, furniture) and interferers (e.g., microwave ovens, cordless phones) are present. This can result in retransmissions that degrade bandwidth and increase latency and jitter. Two basic mechanisms proposed by the IEEE 802.11h Task Group to overcome this problem include Dynamic Channel Selection (DCS) and Transmit Power Control (TPC).

DCS allows a basic service set (BSS) or wireless subnet to be moved to an adjacent channel that may offer better channel characteristics. This preserves QoS metrics and provides a mechanism for overlapping (but different) BSSs (e.g., in adjacent apartments) to avoid one another. TPC enables devices to communicate at the minimum power. In doing so, interference in adjacent or overlapping BSSs using the same channel is reduced, thereby improving the level of QoS for all BSSs.

4.4.3 IEEE 802.11i

Although home users may not be as concerned about security for applications such as surfing the Web, several security mechanisms for wireless home networks can still be adopted. The IEEE 802.11i Task Group has proposed two new encryption schemes:

- Wired Equivalent Privacy 2 (WEP2) using 128-bit keys
- Advanced Encryption Standard (AES)

A secure procedure based on 802.1x for automatically allocating and distributing encryption keys has also been defined as discussed in Chapter 3.

4.5   Case Study: Integrating 802.11 and Hybrid Fiber-Coax (HFC) Cable Networks

Two-way last-mile access technologies that leverage existing infrastructure—digital subscriber line (DSL) and hybrid fiber–coax (HFC)—are maturing; others such as fixed wireless, powerline, and fiber-to-the-home (FTTH) are being developed and deployed on a smaller scale [7]. A HFC cable network comprising cable modems (CMs) and cable modem termination systems (CMTSs) adopts the Data-Over-Cable Service Interface Specification (DOCSIS™) 1.1, which is currently the only last-mile standard with QoS ratified. The latest standard, DOCSIS 2.0, offers true bidirectional broadband support and makes it possible for cable companies to offer end-to-end integrated services and new applications and features such as multiuser gaming and music file uploading.

Cable operators have an advantage over other last-mile access counterparts when it comes to broadband services since they have vast experience in dealing with bandwidth-intensive video traffic. They also have a large subscriber base since TV-centric services, unlike the Internet, draw a wider audience of both young and old. Although less pervasive than Wi-Fi (cable technologies are targeted primarily for the residential and small business market), the subscriber's CM is a versatile device that can potentially integrate video, voice, and Internet access over a single device, providing unparalleled broadband services.

4.5.1 Ongoing Initiatives

At CableLabs® [12], an initiative called PacketCable™ has been tasked to develop specifications for delivering packet services (e.g., voice, video, and other multimedia) over cable networks using IP. Much of the work has been incorporated into the DOCSIS 1.1 specification. The initial motivation for this project was to replace traditional circuit-switched telephone service, enabling competitive local exchange carriers (CLECs) to break into the telephone business. The key specifications are 7 QoS signaling interfaces that support Internet telephony. The scope of work is expected to expand beyond voice services to enhanced multimedia services such as multiplayer gaming and video-conferencing.

CableHome is a more recent program aimed at providing in-home networking and broadband distribution features for connecting DOCSIS devices. The CableHome™ 1.0 specification released in April 2002 offers firewall, Dynamic Host Configuration Protocol (DHCP), and other LAN software features. At the beginning of 2002, CableLabs combined PacketCable, CableHome, and DOCSIS into a unified program called Broadband Access. More information can be found in [8, 9, 11].

4.5.2 An Integrated 802.11/DOCSIS Architecture

The commercial success of the 802.11 and DOCSIS standards provide a compelling motivation for studying how these standards can be integrated in the best possible way. The current practice of most vendors is to implement the two standards independently on a single device, which is shown in the following sections to be far from optimal.

The overall architecture and protocol stacks are illustrated in Figure 4.4. Note that since 802.11 and DOCSIS both support Ethernet packet transmission, end-to-end Ethernet connectivity is possible, which simplifies IP deployment considerably. The following sections identify and address QoS issues pertaining to the integrated standards. These issues are in addition to the ones discussed in Sections 4.2 and 4.3.

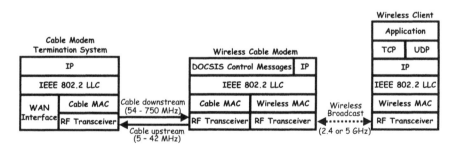

Figure 4.4: Integrated Protocol Stacks.

4.5.3 Integrated Scheduling and Fragmentation at the MAC Layer

The 802.11 and DOCSIS MAC protocols can be tightly integrated for optimum performance. This can be achieved using a combined scheduling mechanism that maps the packet transmission intervals of DOCSIS and 802.11 MAC protocols after the incoming wireless packet has been classified and forwarded to the correct priority queue.

Closely associated with this integration is the issue of packet fragmentation at the respective MAC layers when servicing real-time but low-rate traffic (e.g., Internet telephony) in the presence of high-rate, non-real-time data applications (e.g., Web browsing). The upstream fragmentation feature for CMs can be enabled on a per-service flow basis by the CMTS, providing an important transport mechanism for interleaving large chunks of data packets with low-rate voice traffic. The current 802.11 standard also supports fragmentation at the MAC layer, although the main purpose is to mitigate the high error probability of transmitting long packets over the wireless medium. Clearly, optimizing fragmentation lengths and managing the associated overheads for the two standards can have far-reaching implications.

4.5.4 Throughput Matching

For wireless CMs, special attention needs to be focused on the interface at the aggregation point where there is a capacity mismatch between wireless and cable connections. As explained in Section 2.13, the usable 802.11 throughput varies from as low as 40% to about 80% of the raw rates. Like 802.11, DOCSIS also specifies multiple data rates. The raw data rate for each 6 MHz downstream channel ranges from 30.34 to 42.88 Mbit/s, with usable throughput ranging from 26.97 to 38.81 Mbit/s, respectively (efficiencies of about 90%, which are in contrast with the 802.11 efficiencies). The upstream raw rate ranges from 0.32 Mbit/s to 5.12 Mbit/s (upgradeable to 10.24 Mbit/s). These rates represent the maximum possible since no contention sharing among multiple users is assumed. Clearly, constructing and managing buffers at the wireless CM to minimize the probability of packet loss and delay jitter will be a fertile ground for research, particularly when multiple users are involved.

4.5.5 Network Security and Privacy

A standalone wireless network is prone to intrusions and malicious attacks, but these vulnerabilities can be made worse when coupled with the always-on feature of cable networks. This is because the applications and operating system on the home computer are prone to security defects and virus infection, making uninvited access easy. Some cable operators are concerned about hacker liability and wireless home networks that can give broadband access to and entire neighborhoods, are unsure about the impact of this technology on their business. The encryption schemes in the 802.11 and DOCSIS standards are unfortunately not compatible. As such, higher-layer countermeasures involving firewalls, intrusion detection, 802.1x, and end-to-end virtual private networking (VPN), as depicted in Figure 4.5, are more attractive solutions. VPNs can remove security gaps for end-to-end IP packet transmission among different networks (wired and wireless), thereby providing an integrated security system.

To service IP telephony or multimedia traffic, VPNs using encrypted IP security (IPsec) tunnels may have an advantage over firewalls. In the firewall approach, the firewall needs to act as a proxy for signaling protocols such as the widely deployed H.323 protocol and the emerging Session Initiation Protocol (SIP). Without this proxy feature, the firewall will need to open ports to allow voice connections. This is done without determining whether packets are legitimate, thereby exposing the network to intruders who may spoof a firewall and gain access using voice traffic. However, the proxy method is not without disadvantages. Network address translation (NAT) requires a firewall to parse and modify the contents of a signaling packet all the way up to the application layer. Since NAT changes the source IP address of a packet from private to public to allow it to be routed over the Internet, the firewall performing NAT must keep track of the private IP address in order for reverse traffic to be routed to the sending device.

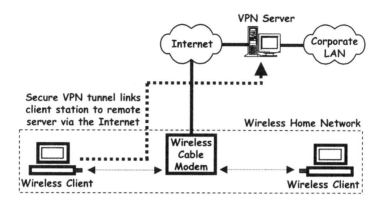

Figure 4.5: VPN Tunneling Removes Security Gaps.

4.6   Summary

This chapter has provided a broad survey of the issues related to QoS provisioning in integrated home wireless networks. It also covered a technical case study involving the integration of 802.11 and DOCSIS standards. We began by covering some qualitative issues on QoS provisioning, and then identified specific problems when supporting disparate traffic types, some of which are open research problems. We then discussed the major QoS features proposed by 802.11e. Finally, we presented an architectural framework, in terms of packet scheduling, fragmentation, throughput, and security of the integrated systems, that will set the directions for future research studies.

References

[1] Benny Bing, *High-Speed Wireless ATM and LANs*, Artech House, 2000.

[2] Benny Bing (ed.), *Wireless Local Area Networks: The New Wireless Revolution*, Wiley, 2002.

[3] A. Dutta-Roy, "The Cost of Quality in Internet-Style Networks," *IEEE Spectrum*, Vol. 37, No. 9, September 2000, pp. 57 – 62.

[4] J. Fijoleck, et al., *Cable Modems: Current Technologies and Applications*, IEC-IEEE Press, 1999.

[5] Zygmunt Haas, "Design Methodologies for Adaptive Multimedia Networks," *IEEE Communications Magazine*, Vol. 39, No. 11, November 2001, pp. 106 – 107.

[6] D. Hartman and T. Quigley, "Multimedia Cable-Networks System Media Access-Control Protocol Performance Simulation," appearing in [4].

[7] Nikil Jayant, et al., *Broadband: Bringing Home the Bits*, National Academy Press, 2002.

[8] Mark Laubach, David Farber, and Stephen Dukes, *Delivering Internet Connections Over Cable*, Wiley, 2001.

[9] E. Miller, F. Andreasen, and G. Russell, "The PacketCable Architecture," *IEEE Communications Magazine,* Vol. 39, No. 6, June 2001, pp. 90 – 96.

[10] P. Wallich, "Digital Hubbub," *IEEE Spectrum*, Vol. 39, No. 7, July 2002, pp. 26 – 31.

[11] L. Wirbel, "The Changing Face of Cable Modems," *Communication Systems Design*, Vol. 8, No. 7, July 2002, pp. 12 – 18.

[12] CableLabs Web Site, http://www.cablelabs.org.

[13] IEEE 802.11 Task Group E. Supplement to IEEE Std 802.11 – Part 11: Wireless Medium Access Control (MAC) and Physical Layer Specifications (PHY): Medium Access Control (MAC) Enhancements for Quality of Service (QoS). Technical Report, IEEE Standards Department, Draft 4.0, Nov 2002.

Chapter 5

Wi-Fi Hotspots

There are currently a few thousand Wi-Fi-enabled locations all over the world, offering fast commercial Internet and secure corporate network connectivity in so-called public "hotspots" namely, major airports and hotel chains, restaurants, cafés, convention centers, and other places where people can rest and relax. These hotspot areas provide an easy and affordable way for business travelers to go online without the inconvenience of fussing over wires and scouting out a network port or phone jack. The increasing attractiveness of such public access networks is poised to launch another rich avenue of growth in the future. According to Cahners In-Stat, public-area wireless LAN revenue will reach $747 million by 2005.

Even with the presence of these local franchise hotspots offering broadband Internet access, wireless LANs are still limited by their roaming capability and user management features. These weaknesses can be overcome by combining high-speed wireless LANs with the large-scale public infrastructure of mobile cellphone networks under a unified billing/identification system. This gives the cellphone operator a major competitive advantage over current hotspot providers who have neither a large mobile customer base nor a cellphone-type roaming service.

5.1   Enabling Technologies

There are several enabling technologies that sparked the recent demand for hotspot services. First, Wi-Fi network adapters are no longer add-ons involving PC cards; with newer versions of notebook computers and other mobile devices are starting to have such adapters integrated into them. In addition to this convenience, a Wi-Fi user can establish communication with access points built by different manufacturers because the adapters are certified for device-level multivendor interoperability by the WFA. Further progress toward large-scale roaming is driven by another WFA initiative called Wireless Internet Service Provide Roaming (WISPr) that allows the subscriber of one WISP to roam to another. Finally, Microsoft has embedded Wi-Fi capabilities into the Windows XP operating system that automatically searches for a Wi-Fi access point, allowing a user to locate hotspot service easily. Microsoft is also partnering with T-Mobile, Boingo Wireless, and Wayport to promote hotspot usage in the U.S. in its new Windows Mobile Pocket PC operating system.

5.2   Advantages and Disadvantages for Deploying Hotspots

There are numerous reasons for deploying a wireless LAN hotspot:

- The high bandwidth-to-cost ratio makes it cost-effective to deploy high-speed broadband services and allows cellphone operators to complement 3G/2.5G technologies.

- Deploying such a service helps to attract new corporate customers and retain current ones.

- Cellphone operators can simplify the partnership with owners of hotspots.

- There has been an increasing number of mobile Internet users, with Goldman Sachs projecting that there will be more than 95 million such users by 2004.

However, public-area use represents only a fraction of the fast-growing Wi-Fi business market. The industry needs to overcome several challenges in making public space Wi-Fi service ubiquitous, including:

- Discontinuous services offered primarily by fragmented hotspot operators (HSOs) and little-known wireless ISPs (WISPs) such as Joltage, AirPath, Concourse Communications, and Surf and Sip. Partnership with many owners of hotspots is needed to achieve a critical mass of access zones.

- Lack of unified roaming among different WISPs forcing users to subscribe to individual WISPs in each area they frequent. This is not only inconvenient but also difficult for the corporate network group to track usage and billing.

- Unproven business models in which property owners and WISPs both want control and share of the revenue. As an example of the business risk involved, MobileStar Network filed for bankruptcy after enabling hundreds of Starbucks coffee shops and other properties with Wi-Fi service.

- Wi-Fi itself is struggling to overcome technical problems, most notably weak security.

Until the number of hotspots reaches a critical mass and roaming agreements are in place, it will be difficult for the technology to become widely accepted. Since it is unclear which strategy will succeed in this embryonic business, many Wi-Fi service entrepreneurs are simply staking out real estate, locking up prime hotspot sites for potential future revenue [4].

5.3   Wireless ISP Roaming

Unlike cellphone networks, which have a much larger reach (coverage area) and subscriber base than wireless LANs, it can be cost-prohibitive for a single HSO to provide ubiquitous hotspot services. On the other hand, by having an ad-hoc patchwork of Wi-Fi locations and different ways to authenticate themselves and log on is confusing and difficult for users.

Currently, two possibilities are offered to carriers to interconnect their wireless LANs:

- Peer agreements: Two operators establish a link between their authentication, accounting, and authentication (AAA) servers. These bilateral agreements already exist for GSM and GPRS roaming.

- Internet roaming: Some companies are today providing global Internet roaming, allowing seamless inter-connections and financial settlements.

Wi-Fi HSOs are nearing agreement on an open, standardized system that will not only allow end-users (mostly business travelers) to access the Internet or corporate intranets, from any of the providers' hotspots but also allow providers to sort out the revenue when a customer roams from one WISP territory to another. WISPr, which was launched by the WFA in early 2001, aims to make public wireless LAN services behave like cellphone services with a common mechanism for handling AAA functions. The project was completed in late October 2002 with a WFA document that specifies the steps needed for a HSO to create a simple, consistent user logon and a way for users to log on to a WISPr wireless LAN with a Wi-Fi-enabled device and a Web browser. The recommendations also included a set of attributes to be used in configuring authentication servers and databases using the RADIUS protocol. Several optional recommendations allow higher levels of security such as the use of gateways between the access point and the provider's network. Another optional scheme allows public key cryptography to be incorporated into the Wi-Fi public system.

Starting in 2003, corporate users will be able to log on at any public access hotspot identified by the new WFA logo—Wi-Fi Zone. These users will be able to gain network access via the nearest provider regardless of where they are and yet have all charges appear on one consolidated bill from the home provider.

5.4   WISPr Operation

The visited RADIUS server relays authentication information to the home RADIUS server, which can either perform the user's authentication or relay the information to the corporate RADIUS server. If authentication is successful, the visited access server assigns an IP address to the user so that he can become part of the visited network. Interactions between the RADIUS servers can be protected using IPSec or Secure Sockets Layer (SSL). Figure 5.1 summarizes the steps involved:

- User logs on to the Web using a username and password (username follows a standardized format).

- User is authenticated by RADIUS servers located at participating HSOs (some HSOs employ secondary authentication methods such as HTTPS).

- User is given access to corporate intranets or the Internet with billing handled by the HSO.

5.5   Components of Wi-Fi Hotspot Chain

There are typically three key components in the hotspot chain—the property owner, the network provider, and the aggregator. The property owner provides the location for housing the Wi-Fi network. Owners such as hotels or corporations can incur the cost of deployment and management to have a franchise Wi-Fi network installed on their premises but receive 100% of the revenue. Alternatively, owners can either collect rent from network providers who wish to use their premises or broker a deal with these providers to share the revenue from the Wi-Fi service.

An aggregator helps to bring in more customers to use the Wi-Fi service, usually on a national or global basis. They also provide enhanced services, centralized billing, and technical support. Carriers have an advantage as they do not need to modify their existing billing system. Typically, a separate rating engine collects Call Data Records (CDRs) from the charging gateway. The rating engine then sends the rated file to the existing billing system. This can be upgraded to collect wireless LAN CDRs from a RADIUS server. Unlike the charging gateway, the RADIUS server issues billing information based on time usage.

5.6   Wi-Fi ISPs

Many Wi-Fi ISPs have set up Wi-Fi hotspots in the U.S. by reselling wireless Internet access across the country with little overhead. The leading Wi-Fi ISPs, Boingo™ Wireless and Wayport, target business travelers.

Instead of building networks and owning infrastructure, Boingo Wireless (which has Sprint PCS as an investor) adopts a more cost-effective approach by aggregating the existing networks of various HSOs together into a nationwide hotspot network. As an aggregator, Boingo provides marketing, technical support, end-user software, and billing while the HSOs build and operate their Wi-Fi networks. Boingo also makes it easy for coffee-shop owners and other venue operators to deploy their own hotspots and join the Boingo network. Boingo helps mobile users with Wi-Fi cards locate nearby wireless LANs using software that can be downloaded free from the company's website. The software provides users a searchable database of access points (over 800 hotspot locations covering 300 cities and 43 states) that is updated daily and allows users to identify Boingo-enabled locations without being on the Internet. The software also provides simple tools for managing multiple network profiles, WEP keys, as well as secure connections using VPN. Boingo charges users from $20 to up to $75 a month to connect to its hotspot service.

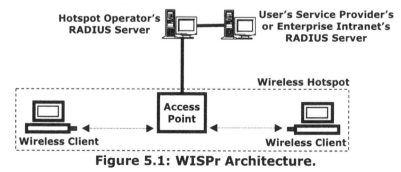

Figure 5.1: WISPr Architecture.

Wayport boasts several hundred public Wi-Fi hotspots, located primarily in hotels. Wayport's service is offered to more than 525 hotels, including the Four Seasons, Wyndham, Hilton, Marriott, Sheraton, Doubletree, and Embassy Suites chains. Verizon Communications has converted 150 of its payphones in New York into Wi-Fi access points and plans to increase the number to 1,000 by end-2003. Verizon's DSL subscribers may use the hotspots at no additional cost. Similarly, Bell-Canada has created 300-foot hotspot zones around payphones in high-traffic locations such as train stations, public squares, convention centers, and corporate campuses.

In Europe, Telia Mobile has deployed over 100 hotspots in Sweden and Norway but charges slightly higher fees. In France, Italy, and the U.K., public networks are not permitted. In some other countries such as Belgium and Germany, a license is required for public usage. Intel has helped implement 802.11 hotspots in several European airports—London Heathrow, London Gatwick, London Stansted, Aberdeen (Scotland), Paris Charles de Gaulle, and Frankfurt (Germany). Charges for access typically costs $10 per hour. Roaming agreements among these hotspots are few, and such agreements are typically managed by phone carriers British Telecom and Swisscom.

5.7 Wi-Fi/Cellphone Network Integration

Combining the reach and mobility features of wide-area cellphone networks with the high data rate of wireless LANs is made possible by the recent evolution of voice-based cellphone networks to data-centric, Internet-based networks. For example, GSM's Short Message Service (SMS) has been very successful in recent years, and an improved version of GSM, the General Packet Radio Service (GPRS) allows voice and data transmission as well as mobile Internet access. Such integration allows the possibility of always on communication through the use of vertical handoffs. In this case, a client uses the cellphone network for default connectivity while scanning for Wi-Fi access points. When an access point is found, the client switches to the Wi-Fi connection.

The level of integration gives rise to several coupling methods. A tight coupling method offers the important advantage of continuous service between the two networks but requires investment on specific access equipment and standardization efforts for new network interfaces. An open coupling method is probably the easiest to deploy since existing equipment can be used, but suffers the disadvantage of discontinuous service. A loose coupling model is an intermediate model between the tight and open coupling methods. In this case, the combined cellphone and wireless LAN systems more or less operate independently, each network type retaining its unique characteristics for ease of deployment and less dependence on evolving standards. Unlike the open coupling method, both the tight and loose coupling methods offer tighter security through the use of SIM cards.

Nokia recently offered a PC card with both Wi-Fi and GPRS cellular capabilities whereas companies such as Ericsson and Lucent have started shipping integrated Wi-Fi cellular products. In the next section, an architecture for cellphone/wireless LAN integration is presented.

5.8   Public Wireless LAN for Mobile Operators— Wireless LAN Beyond the Enterprise (Case Study by Philippe Laine, CTO/NSG, Alcatel)

Wireless LANs find their origins in the enterprise. Yet wireless LAN applications in the public environment are on the rise, particularly in areas of limited physical scope such as airports, hotels, and conference centers. Public wireless LANs can hardly be seen as competing with true mobile data systems. However, they can be deployed as a complementary service to GPRS/UMTS, owing essentially to their bandwidth:cost ratio. Indeed, simple public wireless LANs could be deployed today. On the other hand, they are the subject of substantial development effort, with the result that public wireless LANs are likely to evolve considerably in the coming year or so. This case study presents Alcatel's wireless LAN solution for mobile operators, highlighting key features such as security, session continuity, roaming, and billing.

5.8.1   Introduction

The wireless LAN was originally designed for use inside corporate premises. Meanwhile, new applications are under development for the residential market (for example, xDSL routers combined with wireless LAN) and the public environment. Some applications of wireless LANs in the public environment include:

- Local Independent Service Providers use wireless LAN technology to offer broadband Internet access to rural areas and small cities.

- Community services (for example, Seattle Wireless, NYC Wireless, Consume, and others) offer wireless LAN access to members of the community. These services are managed directly by members of the community.

- Commercial services such as Wayport, BTOpenZone, or HomeRun propose nationwide wireless LAN access in major airports and hotel chains.

This latter service is the only one that is complementary to mobile operator offerings. The others are more comparable to private access.

5.8.2   Choosing Terminals for Public Wireless LAN Access

Any laptop PC can be endowed with wireless LAN connectivity simply by adding a PC card. Such cards are readily available for less than US$50. Some laptops already have wireless LAN connectivity built in, and according to PC manufacturers, more and more laptops will hit the market with built-in wireless LAN.

Pocket PC PDAs also offer wireless LAN connectivity through a PC or Compact Flash card. Some PDAs are even available with built-in GSM/GPRS and wireless LAN connectivity allowing GPRS and wireless LAN access. The Palm OS PDA, after a delayed market launch, now also offers wireless LAN connectivity with a specific add-on. It is likely that smart phones with embedded wireless LAN capabilities will make their debut in a few years' time.

5.8.3   Public Wireless LAN Market Forecasts

In a recent study, In-Stat forecasts a fairly consistent growth of public wireless LAN hotspots, particularly in Europe and Asia/Pacific (Table 5.1). Growth will be strongest in early years, as the market is being developed, since mobile operators will want to be the first to market and attract and solidify their user base. There were approximately 8,000 hotspots worldwide at the end of 2002. Although non-telecom companies own the majority of these public hotspots today, In-Stat forecasts that the mobile operators will bill most of the wireless LAN subscribers by 2006.

5.8.4   The Business Model for Mobile Operators

Hotspots usually belong to companies such as airport authorities or hotel chains. It makes sense for these companies to deploy a wireless LAN, to use it for their own applications, and to sell the unused capacity to telecommunications operators. These companies can directly manage installation and operation, or they can ask an intermediary wholesale access provider to do it for them (Figure 5.2).

Table 5.1:   Market Forecasts for Public Wireless LAN Usage (Source: Alcatel)

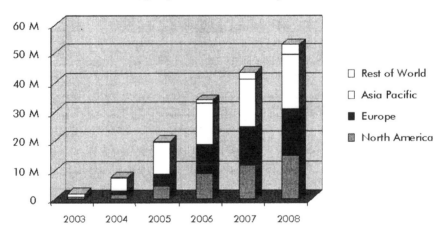

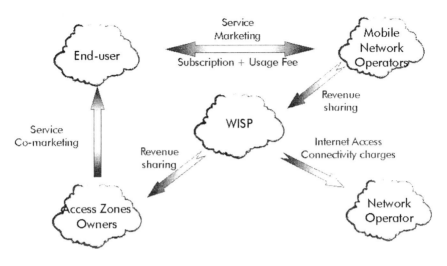

Figure 5.2: **Wireless LAN Business Model for Mobile Operators.**

In this model, mobile operators do not deploy the access network. Rather, they buy capacity on networks deployed by others and manage customer care and billing. The end-users will pay their mobile operator for the usage of the wireless LAN. For the mobile operator, another option is to deploy its own wireless LAN to facilitate end-to-end operation. The option shown in Figure 5.3 illustrates that mobile operators could establish a partnership with a WISP to market their services. For mobile operators, synergy with current data service is essential. This can be achieved by way of a bundled offering (GPRS + wireless LAN).

5.8.5 Public Wireless LAN Applications

In the first phase, public wireless LANs will be used for nomadic usage (On the Pause). The end-user will sit in a hotspot, open his/her laptop and expect to obtain easy access to the Internet, the company Intranet, or his/her e-mail account, etc.; in short, a kind of mobile office. End-users will probably also expect to have access to local information (flight departure in an airport, restaurant menus in a hotel, conference schedule in a conference center, and so on).

	Messaging	**Entertainment and Information**	**Mobile Office**
At Home > Bandwidth > Usage easiness	**E-mail, Instant Messaging**	**Video and Audio** streaming, rich media and software download, gaming, ...	**Intranet,** Internet, File Transfer
On the Move > Time critical > Local info. > Usage easiness	**MMS, Instant Messaging, E-mail** checking, ...	**Video and Audio** clip (extrad), Ring-tones, Gaming, News, Alerts, **Local Information,** ...	**Mobile Office,** Vertical Applications, Itineraries, Schedule, Reservation, ...
On the Pause > Bandwidth > Local Info. > Usage easiness	**E-mail, Instant Messaging**	**Video & Audio** streaming, Rich media & software download, Gaming, **Local Information**	**Intranet,** Internet, File Transfer, **Local Information**
At the Office > Bandwidth > Usage easiness	**E-mail, Instant Messaging**		**Intranet,** Internet, File Transfer

Figure 5.3: Applications in Various Environments.

A second phase will see the seamless availability of applications from mobile and wireless LAN networks. Local services linked to the hotspot locations will also be implemented to offer end users specific applications such as video streaming, infotainment, etc. In addition, services specifically aimed at wireless LAN users could be deployed via mobile networks—for example, the communication of wireless LAN service availability in the area.

5.8.6   Alcatel Solution for Mobile Operators

The so-called "symbiotic relationship" between wireless LAN access technology and mobile operators becomes a reality by coupling the wireless LAN with the mobile operator's core network. The Alcatel solution (Figure 5.4) is flexible enough to be tailored to:

- Operators who wish to interface existing or third party hotspot equipment

- Operators who wish to deploy a complete end-to-end wireless LAN solution from the hotspot equipment to the mobile network

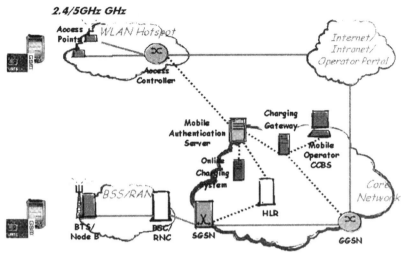

Figure 5.4: Alcatel Solution.

Furthermore, the solution can adapt to various conditions such as:

- Hotspot size and the volume of generated traffic
- Type of existing equipment in the core network
- Security requirements
- Public wireless LAN market uptake

The Alcatel solution allows operators to deploy an end-to-end wireless LAN solution by providing:

- *Public wireless LAN Access Equipment* specifically designed to combine the functions of an Access Point and an Access Controller into a single device

- *Mobile Network Equipment* tailored for mobile operators, such as the Home Location Register (HLR), the Charging Gateway, the online charging system, and the Mobile Authentication Server. This server acts as a gateway between the wireless LAN access network and the mobile network.

- *Backhaul Networks* allowing various types of connectivity between the wireless LAN access network and the mobile operator's network such as xDSL, Frame Relay, Leased Line, and LMDS

5.8.7   Key Features of the Alcatel Solution

5.8.7.1 Security

Two different issues should be considered: access authentication and data transfer confidentiality. For access authentication, there are two possible options:

- *IEEE 802.1x (EAP) authentication* that reuses the SIM card mechanisms. In this case, 802.1x supplies credentials to the Home HLR through the AAA server(s). The SIM card could be a "virtual" if the terminal is not equipped with a SIM card reader.

- *Secured WEB-based authentication* that uses a username/password pair submitted to the Home AAA server. This password could be a "one-time password" delivered on a scratch card or via SMS to the customer's cellphone.

SIM authentication offers more security and should be preferred by mobile operators but it requires client software installed on the customer terminal. Both solutions should be supported simultaneously, in particular for roamers from WISP or fixed-operator networks (Figures 5.5 and 5.6). Once the user is authenticated and authorized, user data needs to be protected from interception. Not all data is, of course, considered confidential, but critical information should not be transmitted unless security measures have been taken. The solution for data transfer confidentiality allows adopting data protection at different layers:

- At the *Data Link Layer*: communication between a wireless device and access point in the air interface is kept private using the Wired Equivalent Privacy (WEP) protocol. However, as it currently stands, WEP has been shown to contain flaws and does not meet the claimed security level. Moreover, the lack of a key management protocol raises major concerns (e.g., a secret key is chosen manually). Mechanisms for dynamically assigning and renewing WEP keys are required.

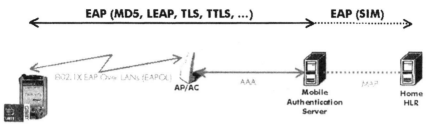

Figure 5.5: SIM Card Authentication.

- At the *Network Layer*: communication between the wireless device and the corporate network is kept private using Virtual Private Network (VPN). An immediate solution to build secure communication is to make use of an IPSec-based VPN, at least when setting up wireless LAN access to corporate networks. Having established IP connectivity, the end-user is authenticated when setting up the IPSec tunnel to the corporate network. Terminals must embed an IPSec client to be able to set up the VPN.

- At the *Application Layer*: communication between a wireless device and an application server such as a critical web server can be further protected using https/SSL.

Later on, new mechanisms will be made available to provide a secure protocol at the wireless network level. The IEEE 802.11i Task Group is currently studying solutions. The interim solution will be the Temporal Key Integrity Protocol (TKIP) that effectively makes the WEP key dynamic. The long-term solution will be based on the Advanced Encryption Standard (AES), which has a far better underlying cipher.

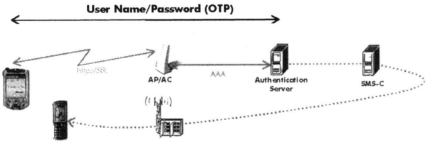

Figure 5.6: One-Time Password Authentication.

5.8.7.2 Session Continuity

Session continuity addresses the change of access network while a session is ongoing. The change of access network can be between wireless LAN and wireless LAN or between wireless LAN and 2.5/3G networks. The mobility is merely coordinated at the IP level using Mobile IP (MIP) allowing a tolerable interruption and recovery of the ongoing session. It is not a conventional network handover. The typical network architecture recommended by 3GPP and IETF requires:

- Support of Foreign Agent (FA) functionality in the mobile network. Typically, the FA would reside in the GGSN.

- Support of FA in the wireless LAN access network. Typically, the FA would reside in the Access Controller.

- Home Agent (HA) can be anywhere (PLMN, ISP, Corporate network, etc.).

The Alcatel solution relies on an intelligent MIP client that can even accommodate access networks with no FA support. When no FA functionality is available in the visited network, the client acquires a Co-located Care of Address and tunnelling is done all the way to the terminal.

5.8.7.3 Roaming

The Alcatel solution supports two roaming methods:

- *IP roaming*, which makes use of the AAA mechanisms
- *GSM roaming* using SS7 mechanisms

5.8.7.4 IP roaming

IP Roaming is supported thanks to the inherent AAA mechanisms. The AAA server of the visited network forwards the authentication request to the home AAA server (in this case, the Mobile Authentication Server). This request can go through a chain of AAA proxies before it reaches the home network. The Mobile Authentication Server authenticates the user, and accesses the HLR for cellular-grade authentication. If the authentication in the home network is successful, the AAA in the visited network is instructed to enable customer session. The visited network AAA sends accounting information to the home network (Figure 5.7).

It is to be noted that IP roaming was adopted by 3GPP as the fundamental principle for roaming. By relying on IP roaming, mobile operators can support roaming not only with other mobile operators, but also with WISP and fixed operators (Figure 5.8).

5.8.7.5 GSM Roaming

As the Mobile Authentication Server supports standard SS7 and MAP protocols, it provides flexibility for mobile operators wishing to rely on GSM roaming with other mobile operators. In this case, the Mobile Authentication Server located in the visited network communicates with the HLR/AuC of the home network for authentication.

Alcatel recommends adopting the IP roaming mechanism promoted by the 3GPP. However, GSM roaming is also part of the Alcatel solution.

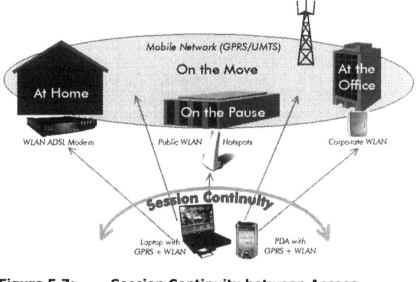

Figure 5.7:   Session Continuity between Access Networks.

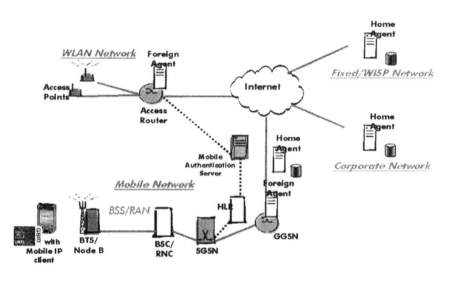

Figure 5.8: Mobile IP Architecture.

5.8.7.6 Billing

The Alcatel Public Wireless LAN billing solution provides several billing schemes for postpaid subscribers:

• *Volume-based billing:* The subscriber is billed according to volume of data uploaded/downloaded.

• *Duration-based billing:* The subscriber is billed according to the duration of the connection.

• *Flat-rate billing:* The subscriber pays a fixed subscription for unlimited usage. This is an "all you can eat" type of billing.

• *No billing:* In some cases, it is useful to provide some access to some services for free, for example, to get local information at the airport such as flight or duty free information.

• *Differentiated billing:* The subscriber pays for content, for example, to view a movie while waiting at the airport.

The wireless LAN access network issues the relevant accounting information to the AAA server. Accounting information could be incorporated into operator's specific CDRs, allowing them to work with existing operators legacy billing systems and to create a single bill GPRS and wireless LAN usage. In addition, an enhanced set of billing schemes can be exploited allowing temporary access or handling of prepaid accounts:

- Vouchers: This is a convenient way for enabling access for temporary account users such as those using scratch cards. This allows occasional users to have public wireless LAN access.

- Prepaid accounts: The subscriber decides on the amount of wireless LAN usage by means of a prepaid account.

5.8.8   Conclusions

Public wireless LANs provide an interesting complement to mobile networks, enabling broadband Internet access in selected hotspots and offering additional capacity. Public wireless LANs can be deployed *now* with a simple and cost-effective solution based on existing equipment. Mobile operators can offer their corporate customers a bundled offering of GPRS plus wireless LAN, managing the subscription to the service as well as customer care and billing. In addition, they can deploy or reuse existing prepaid servers or payment servers as a flexible way of charging customers. Interworking between networks is mandatory for ease of use, the public wireless LAN being a complementary IP access using mobile network security and subscriber management. Seamless continuity between mobile and wireless LAN networks is possible with the introduction of session mobility using Mobile IP.

The wireless LAN, being at the confluence of enterprise, home networking, fixed, and mobile networks, is best placed to enable new services and applications encompassing the business and residential segments. From this perspective, interworking with mobile networks is only the first step toward an optimized service delivery in heterogeneous network environments.

5.9   Summary

Wi-Fi hotspots typically provide a uniform logon and user experience for broadband Internet access in places where business travelers tend to be located. They have started to become pervasive, with some major phone companies turning their payphones into hotspot zones. Even burger chain McDonalds has started a pilot Wi-Fi hotspot service in 10 Manhattan outlets in New York and plans to expand the service to several hundred restaurants by end-2003. In the current pilot program, an hour's worth of access is free with the purchase of a complete meal; otherwise, access costs $3 and is limited to one hour. WFA's WISPr initiative allows roaming users visiting a Wi-Fi hotspot not managed by their own provider to access the Internet and still be billed by their provider. This is achieved using the AAA mechanism in RADIUS servers.

A mixture of local free hotspots and broad-reach commercial hotspots in major public zones can potentially help make hotspots more common. Pricing is fundamentally important and commercial hotspot services should provide value for access charges such as incorporating enhanced security since business travelers will need to access sensitive information.

The chapter concludes with a case study on how hotspots can provide an interesting complement to cellphone networks. Major implementation issues including billing, security, and roaming are discussed. The role assumed by the RADIUS and 802.1x standards in ensuring network security is critical.

References

[1] Eric Brown, "Wireless LANs Go Public", *MIT Technology Review*, June 18, 2001.

[2] http://www.nwfusion.com/wifi/2002/index.html.

[3] "Special Report – Wireless Internet," *Business Week*, April 29, 2002.

[4] Paul Henry and Hui Luo, "Wi-Fi: What's Next," *IEEE Communications Magazine*, December 2002.

[5] Boingo Wireless Web Site, http://www.boingo.com/hso.

[6] Public Access Wireless LAN in Europe: A Technology in Search of a Business Case?, Yankee Group, June 2002.

[7] Public Wireless LAN Service: Mobile Operators Mustn't Miss the Boat, In-Stat MDR, July 2002.

[8] U.S. Communications Technology: Watch out for Wi-Fi, Goldman Sachs, September 2002.

[9] BluePrint Wi-Fi: European Wireless LAN Review, EyeforWireless, November 2002.

[10] Wi-Fi Integrated Circuits, Allied Business Intelligence, January 2003.

[11] Wi-Fi Zone Web Site, http://www.wi-fizone.org.

Chapter 6

Market Segmentation and Analysis

The wireless LAN market is one of the fastest growing markets today. Over the last three years, a very rapid wireless evolution has occurred in the home and public markets. In 1999 and 2000, HomeRF was considered a strong contender for the technology of choice for the home. In 2001, HomeRF dropped out and Wi-Fi became the natural choice for the home. In 2001, the main issue was whether 802.11a or 802.11g would be the technology of choice for the home and the enterprise. In 2002, the wireless home landscape broadened, with a handful of solid wireless technologies showing significant potential for specific applications in the home.

In a market characterized by rapid growth and commoditization, the wireless LAN chipset market was not spared either. The barriers to entry have been significantly reduced since the standards received approval. The IEEE 802.11 chipset market, especially 802.11b, is the most competitive and the most successful. Although Intersil has the lion's share of the 802.11b market, and Atheros almost 100% of the 802.11a market, the emerging multiband 802.11a/b/g (2.4/5-GHz) market has presented additional opportunities for startups and established players like Broadcom, RF Micro Devices, and TI. Uncertainty over which standard to support has caused many vendors to develop multimode (802.11a/b/g) solutions. In addition, some of the 802.11 chipmakers are also integrating Bluetooth in their designs. This rapid growth, especially in the 802.11b market, has led to rapid hardware and chipset commoditization, with very little margin for both wireless LAN IC manufacturers and hardware vendors. According to Instat-MDR, end user revenues increased only 8%, one-third that of shipment growth, accounting for $532 million for the third quarter of 2002. This has led to shifts in the marketplace, such as that of Cisco exiting the 802.11 client market amid stiff competition from Taiwan Inc. Interestingly, Cisco also made an acquisition in March 2003. They acquired Linksys for $500 million. While the amount may not appear big (1.1 times sales), the shift in Cisco's strategy is more important. What is the bottom line? There is none as far as the pervasiveness of wireless LANs is concerned. The market is just expanding beyond the traditional computing arena. We should expect to see significant activities in the public and home spaces in the coming years. Figure 6.1 sums up nicely the pervasiveness of wireless LANs.

6.1   Major Events in 2002

An increasingly competitive market in both the hardware and chipset arenas marked 2002. The price erosion in chipset ASPs, wireless NICs, and access points (APs) continue to drive the market toward further commoditization. The standards activities have certainly helped to fuel the delivery of newer pre-802.11g products. Pre-802.11g products have entered the market ahead of the 802.11g ratification.

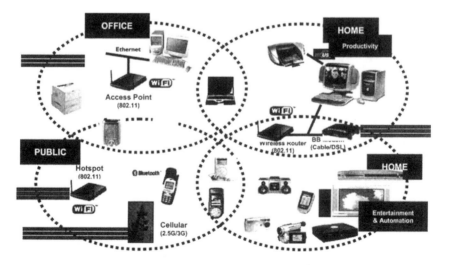

Figure 6.1: The Pervasiveness of Wireless LANs.

The 802.11b market remains the dominant flavor, however. The higher data rate, multi-mode 802.11a, and 802.11a/b/g standards picked up steam in the second half of 2002. Security continued to be an issue toward enterprise adoption in spite of new initiatives undertaken by the IEEE 802.11 Working Group and the WFA.

Market dominance for the 802.11g market will be tightly contested between Intersil and Broadcom. In the multi-mode market, Intersil, Broadcom, and Atheros will battle it out initially. Agere, RF Micro Devices, Philips Semi-conductors, and TI will join them in the latter half of 2003. The greatest threat will potentially come from Intel, with the launch of the Centrino platform. This platform marks the beginning of wireless LAN-on-chipset solutions. Even though they are late to the game, Intel has already pledged millions of marketing dollars to the program and will eventually force notebook manufacturers to move in its direction.

2002 also marked the beginning of a new breed of wireless LAN applications in the form of PDAs, handsets, and digital tablets. In the hardware market, one of the major market developments that took place in 2002 was Proxim's acquisition of the Orinoco range of wireless LAN products from Agere Systems. Although Agere says the sell-off justifies its strategy to focus on its core competency—chipsets and NICs—for Proxim, the acquisition will strengthen its enterprise-class product portfolio. Taiwanese companies like D-Link have exhibited strong growth in 2002, thanks to their aggressive pricing, which has helped penetrate price-sensitive markets. Newer products that emerged in 2002 included an AP switch device from Vivato. Its PacketSteering architecture combined Wi-Fi, smart antennas, and Gigabit Ethernet switching to pave the way for an enterprise class of 802.11 switching architecture. Another area of interest is in the embedded Wi-Fi market. IBM, Toshiba, Fujitsu, Sony, Dell, HP and Gateway all launched notebooks with embedded Wi-Fi in 2002/2003.

There was also growing interest worldwide among small WISPs (Wireless ISPs) as well as large and established carriers in wireless LAN hotspots/hot zones deployments. The business model however has not been fully understood and remains a critical factor for success. This has not stopped companies from deploying Wi-Fi networks in hotels, airport lounges, and coffee shops. According to [3] one of the first mobile carriers to embrace Wi-Fi was VoiceStream Wireless. VoiceStream bought bankrupt MobileStar Networks, which included 1,200 Wi-Fi hotspots at Starbucks locations. VoiceStream expects to have 2,000 hotspots by the end of 2003. In 2003, VoiceStream plans on using the hotspots as part of a rebranded T-Mobile wireless broadband service, allowing its GPRS customers to connect at 11 Mbit/s. For example, Nokia launched a GPRS/wireless LAN PC card allowing subscribers to move from GPRS to Wi-Fi without needing to relogin or reestablish a connection when moving between networks. The seamless movement is possible through a GSM SIM card holding subscriber information. Similarly, Sprint PCS invested $15 million in Boingo Wireless. Boingo Wireless presently has many hotspot locations in cafes; airport lounges; hotels like Marriott, Wyndham, and Four Seasons; and convention centers. Sprint is looking to launch a dual-mode card that allows subscribers to move between hotspots (802.11a/b) and its CDMA network. A start-up company in California, Gtran Wireless, is selling a dual mode CDMA1XRTT/wireless LAN PC card allowing customers to move from a CDMA network to a wireless LAN. The company is reported to have had success in Korea. During Q402, AT&T, Intel, and IBM announced the formation of a joint venture called Cometa Networks to provide a nationwide wireless LAN hotspot network. Cometa is targeting 25–50K access points at various sites including retail stores, restaurants, hotels, and gas stations, with rollout slated to begin Q303 in the top 10 major metropolitan areas. IBM will provide back office services and AT&T will provide the network backbone. Venture capital firms Apax Partners and 3i along with Intel Capital (<$10M commitment) funded the venture. Cometa will offer services to ISPs, telecom carriers, and enterprises. Combined marketing muscle and extensive resources of the syndicate should further accelerate proliferation of wireless LAN adoption, which in turn will benefit chipset suppliers.

Perhaps the best description of 2002 highlights was captured in the In-Stat/MDR December 2002 report [2]. It provided three key highlights, listed in the following sections:

6.1.1   802.11b Brings Low Cost and Value to the Table

2002 marked a tremendous year for Wi-Fi volume growth, buoyed by the maturity of 802.11b products, along with their falling (and thus increasingly attractive) prices. This combination of low cost, value, and increased reliability drove 802.11b the small business, home/SOHO, and, increasingly, into small departments and/or remote offices of large enterprises. Even with 802.11a on the market for a full year, 802.11b had a tremendous third quarter, and in a large sense has pushed vendors to deliver dual-band products, thus allowing increased lifetime of 802.11b products.

6.1.2   Varied Technologies for Different Types of Users

Technologies and equipment continued to evolve, with a-only, dual-band a/b, and even a sprinkling of g coming into the market in 2002. In the enterprise, APs with management and security features over and above the low-end products were in demand from the likes of Cisco, Symbol, 3Com, Avaya, Enterasys, and Proxim.

6.1.3   Wi-Fi Everywhere

Wi-Fi products, especially 802.11b products, took up significant shelf space at retail stores like Best Buy and CompUSA, and were consistently some of the best-selling electronics equipment on mainstream e-tail sites such as Amazon.com and Buy.com, as well as networking equipment sites like cdw.com.

6.2   Trends for 2003

For 2003, the wireless LAN industry will generate US$1.9 billion in total revenue based on industry estimates. Chipsets sales alone are targeted to hit 23 to 25 million units in 2003, up from 7.9 million in 2001. Between 2002 and 2007, Wi-Fi chipset shipments are set to grow at a Compound Annual Growth Rate (CAGR) of 45%. By 2007, shipments will reach 150 million chipsets, with revenues of

US$1.2 billion. There are trends emerging as well in both the hardware and software aspects of a wireless LAN solution. In terms of hardware evolution trends, Table 6.1 illustrates the trends expected for NICs and APs. Finally, the mobile hand-held market will include all types of handsets: cellphones, smart phones, converged phones and PDAs. This last category is expected to ramp up in volume due to the convergence of computing and communications.

Figure 6.2 and Table 6.2 illustrate the concept further. As convergence continues, the opportunities for innovation will increase, driving new capabilities into the marketplace, creating new lifestyle-based usage models, and further spurring demand. This convergence has major implications for the wireless client space and the opportunities it will create across the telecommunications industry, the semiconductor industry, and the entire Internet landscape.

6.3   The Price War Continues

PCMCIA network interface cards (NICs) supporting the 802.11b wireless LAN standard are likely to see a drastic price drop in the second half of 2003, dipping to near US$8 from the current US$13—15, as new players, both chip suppliers and equipment makers, try to get a foothold in the already saturated market. This trend is expected to continue, with 802.11b hitting $5 in the next 18 months.

Table 6.1: AP and NIC Trends for Wi-Fi.

Equipment	1999	2000	2001	2002	2003
NICs	PCMCIA	PCMCIA, USB	PCMCIA, PCI, USB, mini-PCI	PCMCIA, PCI, USB, mini-PCI, CF	PCMCIA, PCI, USB, mini-PCI, CF, SDIO, other embedded stds
APs	Standard	Standard *wireless bb **access	Standard *wireless bb ** access	Standard *wireless bb **access AP/switch	Standard *wireless bb **access AP/switch

*   Wireless bb router = wireless broadband router = wireless gateway = AP/router this device commonly has 3-4 Ethernet ports
**   Access servers: high security APs for public spaces

Table 6.2: Estimated Market Rollout (By Quarter).

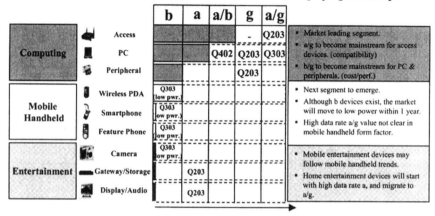

		b	a	a/b	g	a/g	
Computing	Access				-	Q203	• Market leading segment.
	PC			Q402	Q203	Q303	• a/g to become mainstream for access devices. (compatibility)
	Peripheral				Q203		• b/g to become mainstream for PC & peripherals. (cost/perf.)
Mobile Handheld	Wireless PDA	Q303 low pwr.					• Next segment to emerge.
	Smartphone	Q303 low pwr.					• Although b devices exist, the market will move to low power within 1 year.
	Feature Phone	Q303 low pwr.					• High data rate a/g value not clear in mobile handheld form factor.
Entertainment	Camera	Q303 low pwr.					• Mobile entertainment devices may follow mobile handheld trends.
	Gateway/Storage		Q203				• Home entertainment devices will start with high data rate a, and migrate to a/g.
	Display/Audio		Q203				

Bluetooth is already less then $5 now and will be halved in the next 18 months. Companies like Dell expect to replace IrDA on laptops when Bluetooth reaches the $1 price point. There is a very elastic relationship between pricing and demand in the wireless LAN market, and every step down in prices brings the industry closer to unlocking new opportunities and applications. A good example is the expansion from the computing market to the mobile handheld and PDA market. This trend is expected to continue through 2005.

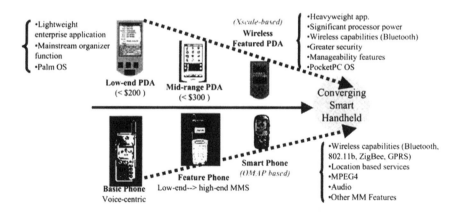

Figure 6.2 : Convergence of Computing/Communication.

6.4   Wireless LAN Market in the Asia-Pacific Region

In 2002, the wireless LAN market in the Asia-Pacific region recovered from the slump that had set in the market in 2001. In 2001, though wireless LAN implementations continued despite the harsh investment conditions, the market did not really take off as expected. The changes in the market were especially evident in the second half of 2002, when the telecom carriers began to move aggressively in the public hotspots space in countries such as Singapore and Australia. Wireless LAN revenues in the Asia-Pacific region are expected to witness some rapid growth in 2003, driven by the services sector. However, the enterprise sector, for which wireless LAN was originally intended as a connectivity option, will also continue to grow as new killer applications such as voice over wireless LAN emerge, and mobile workers begin to extend the usage of wireless LANs beyond simple e-mail and personal information messaging to accessing complex databases and for order entry. Enterprise-focused vendors such as Cisco and Avaya have begun adding secondary layers of security solutions to address the complex needs of enterprises in terms of secure anytime, anywhere access to their corporate networks.

6.4.1   Major Players in the Asia-Pacific Region

In 2002, the maximum concentration of wireless LAN equipment vendors was in Taiwan, which is fast emerging as a popular outsourcing destination for most global vendors. South Korea is another market in which a number of local players compete with multinational vendors. The leading players who have captured significant market share in 2002 are Buffalo (Melco), Cisco, D-link, Avaya, IO-Data, NEC, Corega, Agere, MMC, and Samsung Electro-Mechanics. Between themselves, their revenue accounted for about $154.80 million out of the total revenue of $189.31 million in the first half of 2002. The enterprise-class products roughly accounted for over 70 percent of the market, whereas the SOHO and the retail segment accounted for the rest. The enterprise-class product vendors are moving toward offering end-to-end wireless network solutions rather than just the network, and players who target the home market offer the basic plug-and-play equipment, pricing them very low.

6.4.2   Taiwan – The Next Wireless LAN Powerhouse

In 2002, wireless LAN unit shipments in the Asia-Pacific region rose to 3 million from 1.3 million in 2001, registering a 130 percent growth rate while revenues posted a growth of 58 percent from $266 million to $421 million. The wireless LAN market in the Asia-Pacific region is as diverse as it can get. Broadly categorized into three segments, these will typically be grouped under:

- Early Adopters: Singapore, Japan, South Korea, and Taiwan fall in this category due to the maturity of their wired infrastructure.

- The Promising Followers: Australia, Malaysia, and People's Republic of China are in this group. The general characteristics of this group are their potential to become leaders in the technology due to the vast resources available to them.

- The Laggards: The geographical constraints of the Philippines and the low penetration of wired infrastructure in India have made wireless LAN an interesting alternative to wired infrastructure in these two countries. However, given the low penetration of laptops and devices such as the PDA, the wireless market in India has been restricted to wide-area wireless networks.

It is worth noting that of all the countries mentioned, Taiwan has the potential to lead the next phase of wireless LAN growth. Fueled by the success of the electronics industry, Taiwan's formula for success lies in its innate ability to form cluster industries. The power of this results from a critical mass of players that can help set up a ready client base across a range of sectors. The island's strength in wireless LAN cards and systems as well as the notebook industry further strengthen the growth of the wireless LAN

IC industry. At present count, there are more than 10 local design houses in Taiwan. These companies themselves are backed by Taiwan's base of wireless LAN equipment makers and notebook manufacturers. This powerful electronics ecosystem creates a pent up demand and the intensity of competition is much higher too. A common joke among wireless LAN players is that "you can never keep a secret in Taiwan." Information flow is richly intermingled in a complex web of communication so that the slightest whisper of competitive information is quickly spread through the grapevine. As a practical matter, many U.S.-based companies obtain competitive intelligence through their Taiwanese channels. If past experience is any indicator, Taiwan Inc. has always been able provide the same solution at a much lower price compared to global players. This is in fact the case for the computing market in 802.11b today. Market leader Intersil is losing market share to Taiwan Inc. because the MAC/Baseband and radio technology are becoming common knowledge and further differentiation is becoming increasingly difficult.

Figure 6.3 shows a typical wireless LAN supply map relationship between design houses, ODMs, OEMs, and system integrators.

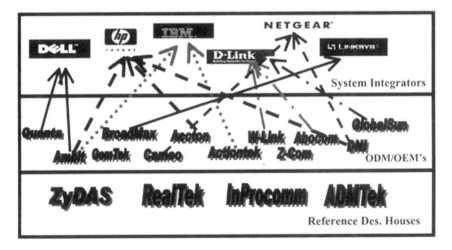

Figure 6.3:   **A Typical Wireless LAN IC Supply Chain for Taiwan.**

Of the wireless LAN IC makers listed in Figure 6.3, some of them such as Inprocomm started in the U.S. Gradually, they were drawn to Taiwan for reasons such as time-to-market, time-to-clients, and also lower costs of engineering and development. Most of the wireless LAN IC makers focus on providing baseband and MAC technology at competitive prices. RF technology is still pretty difficult to master, resulting in many strategic partnerships with U.S. companies such as Philips Semiconductors in order to provide the complete solution.

6.4.3   China: The Wireless Giant Awakens

While the majority of the design activities are happening in Taiwan, its dominant neighbor, China, is already positioning itself as the next wireless LAN powerhouse. Several reports have already suggested that China is likely to take a different path toward wireless LAN proliferation. The trend likely to be adopted is one that mixes a top-down approach from China's big telecom service providers and a bottom-up approach from corner-store owners trying to offer Wi-Fi service. Much like the frenzy that saw cellphones take off in China seven years ago, the prediction is for a similar uptake happening for Wi-Fi adoption.

Currently, the wireless LAN market in China is mainly driven by computing applications. However, there are some companies like Cellon that are working on Cellular and wireless LAN combo cards. It is expected that the market will follow the trend of going from standalone to embedded applications and from 802.11b to 802.11a/b/g. The incumbents like Intersil, Agere, Broadcom, and Atheros provide most of the solutions today. For companies hoping to penetrate this market, it will be helpful to have a multi-prong product, marketing, and financial strategy.

From a carrier perspective, two of China's largest telecom service operators started building wireless LAN networks several years ago. China Netcom, which operates a wired telephone network, and China Mobile, one of China's two wireless phone operators, are offering wireless LAN access in China. China Unicom, which has been busy building a CDMA-based cellular network in China, is also planning to push hard into the Wi-Fi market. Estimates for

Q103 indicate that China has about 2000 Wi-Fi access points, with most of those owned by the large national telecom operators. China Mobile, for example, has perhaps 500,000 to one million customers who use its Wi-Fi access points. China Netcom is promoting the use of prepaid cards, with free trials for early Wi-Fi adopters.

Over the next decade, Greater China (defined as Taiwan, Hong Kong, and mainland China) will be one of the fastest growing regions in the world for major networking technologies. Today there are significant opportunities for wireless LAN, Bluetooth, and USB products in computing, communications, and consumer devices. Major beneficiaries of Wi-Fi in China include local ODMs, local brand name manufacturers, and companies with manufacturing operations/partnerships in greater China.

6.4.3.1 Common Penetration Strategies for China

Partners from other countries such as Taiwan, Korea, and Singapore influence the China market. The market is characterized by low-cost hardware manufacturing moving toward higher-end, value-added service in the near future. In order to have a decent chance of success in China, companies are best served if they have the following arrangements planned:

- Proper design-in support from Taiwan wireless LAN companies that can act as competent centers

- Local support, marketing and sales in China. Technical staff should include field application engineering at major cities/hubs (e.g., Beijing and Shanghai)

- Engage with relevant Chinese organizations to understand regulatory and standards policies

- Build the academic links. In any growing economy, the universities tend to take the lead in the pursuit of knowledge for the nation. Companies should develop links with a prominent technical university to provide access and visibility for the company, its products, and technologies to influential present and future business and technology leaders. Cisco, Lucent, TI, and IBM are good examples of good academic relationships leading to gaining overall market shares in China.

- Leverage the Taiwanese commercial links. Most U.S. multinationals already have commercial links with companies in Taiwan. They usually represent potential high volumes in the communications market segments. Use these partners to cover the Chinese market. In addition, companies can use the current high-volume opportunities to cover China to optimize current resources and avoid spending additional resources and heavy support on low- and medium-term business. For the wireless LAN IC companies, it makes sense to identify and partner with at least two design houses close to the Chinese market. This will enable distribution and add value to the company's solutions. The added value comes from the services provided by design houses, and provided differentiation by using their own upper-layer networking software.

6.5   The Wireless LAN Semiconductor Market

The wireless LAN semiconductor market has become brutal in the face of intense competition and market entry of more players (established and start-ups). As of 2003, there were more than 50 known suppliers of ICs for 802.11 wireless network applications, making a shakeout inevitable. Some of the notable closings include Titan, which closed its LinCom Wireless 802.11 operation, and MicroLinear. The 802.11b-only solution became increasingly commoditized as a result of intense price pressure.

Given the maturity of the standard, it is expected that "b-only" solutions will continue to make up the majority of the market segment, with increasing volumes and rapidly eroding ASPs following the entry of Taiwanese players. The inevitable shakeout will mean that the wireless LAN IC firms that succeed in the next up cycle will be those who identify growing markets and then develop and execute winning plans. Most market research studies indicate that product markets such as the wireless AP semiconductor market are clearly not attractive target segments for the foreseeable future. The wireless LAN markets that do show promise at least in the near term are in consumer electronics and home networking.

A new challenge for wireless LAN IC companies is the entry of Intel into the wireless LAN chipset market. Together with VIA and SiS, they are planning to embed the MAC in South Bridge chipsets in 2003 as wireless LAN-on-chipset products. There are, however, severe technological barriers and will not likely be a threat until 2004. Eventually, this approach will pose a threat to all existing wireless LAN IC suppliers, including Taiwan IC suppliers. Due to the rapid price erosion and the introduction of dual-band chipsets, the market for 802.11a will probably see marginal growth in 2003. According to [1], total unit shipment of 802.11a for 3Q02 was only 105,000 compared to 109,000 units shipped in 2Q02. The fact that there is no easy compatibility with the existing installed base of 802.11b infrastructure further complicates the acceptance. The expected share for 802.11a will be further eroded with the introduction of pre-g products in 2002. Even though these pre-g products from Broadcom and TI have reported interoperability issues, they are expected to be resolved in due course now that 802.11g has been ratified. Most 802.11a-only vendors are shifting their product lines to incorporate dual-mode APs, as end-users are demanding these for future-proofing purposes. They key to 802.11g is the backward compatibility it offers to the existing installed base of 802.11b systems. To take this reasoning further, according to [3], "g-only" solutions are forecasted to account for 30% of the wireless LAN market in 2003. Broadcom's introduction of "g-only" products in late 2002 and the scheduled release of Intersil's Prism GT in Q103 should spur other chipmakers to aggressively push for similar product introductions. Marvell, in February 2003, announced, an 802.11g-version of its Libertas chipsets. Marvell claims they will provide backward compatibility with existing 802.11b designs while offering compatibility with the 802.11g specifications.

Following the lead from Table 6.2 above, one way to dissect the wireless LAN IC market is to focus on three major segments: computing, mobile, and consumer.

6.5.1   The Computing Market

The projected volume for the computing market is shown in Figure 6.4. The salient point from Figure 6.4 is the high adoption of wireless LAN in notebooks. This should not be too surprising given the current market trends. Clearly this bodes well for wireless LAN OEMs like DELL, HP, and IBM. Another significant trend is the integration of wireless LAN capabilities in home networking devices like cable modems, DSL routers, and broadband gateways.

Based on 2002 estimates, 80% of the wireless LAN manufacturing is located in Taiwan. This trend is expected to last as long as commoditization continues and Taiwanese OEMs and ODMs are able to out-price and out-feature U.S. and European companies. Another increasing trend in this market is the transition from 802.11b to 802.11g and 802.11a/g. The drop in bill of materials (BOM) for 802.11a/g products further fuels this trend. Also, with the transition from analog to digital interfaces becoming more commonplace, the baseband and MAC can become part of the PC South Bridge I/O chipset. This leaves only a digital connection to the RF, which will be less susceptible to interference. Some companies are also working on integrating the baseband and MAC with the RF on a single chip for higher performance and lower cost.

One of the key dynamics expected in 2003/2004 is that the price delta for 802.11a/g over 802.11g is expected to erode faster than 802.11g over 802.11b. The critical success factors for thriving in this market for the wireless LAN IC industry include having the lowest BOM cost and fastest time to market. Beyond these two factors, the remaining differentiators include performance and features. For example, the types and level of security compliance with IEEE 802.11i and WPA will be key. Also, the latest Cisco Client Compliant Extension (CCX) should also provide an advantage for those who are certified. The most basic certification is the Wi-Fi certification from the WFA. Microsoft Windows Hardware Quality Labs (WHQL) certification is also an important milestone to achieve. Finally, the level of system integration and driver support can also make a difference between wireless LAN chipset suppliers.

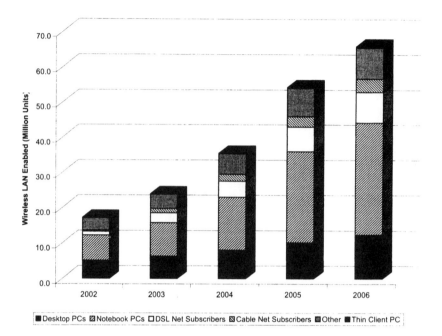

Figure 6.4:   Projections for Computing Market.

6.5.2   The Consumer Market

Two important trends stand out for the consumer electronics industry. The first one is the "WiFi-cation" of devices. If consumer electronics (CE) companies do not have wireless on their roadmap, they are likely going to be edged out by those with Wi-Fi and Bluetooth capabilities.

Fueled by the hype surrounding Intel's Centrino chip, the expectation is for low-power, wireless laptops to be the default configuration coming out of the box. For companies developing such products, it is important to consider security implications not just from the intranet standpoint alone. There are also guests using Internet access in corporate facilities and employees needing to access data from other locations (including places like Starbucks). The projected volume for the consumer market is shown in Figure 6.5.

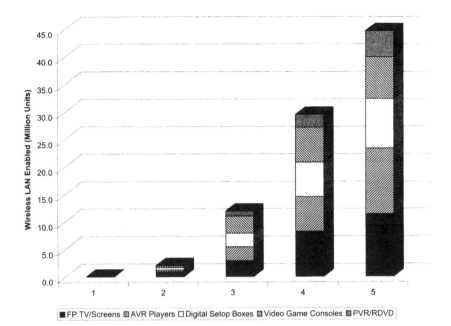

Figure 6.5: Projections for Consumer Market.

The increase in multimedia client equipment to distribute audio-visual information is expected to increase in the home after 2003. This segment of the wireless LAN equipment industry is expected to grow from 7.3 million node shipments in 2001 to 160 million node shipments in 2004. New access points, including cable and DSL modem broadband router gateways and set top boxes, will contribute to this growth in the home.

At the CES 2003 show, Microsoft announced its new Windows-powered Smart Displays. The Smart Display includes a processor, an 802.11b radio, memory, and an embedded version of Windows CE that communicates with a Windows XP machine using its native Remote Desktop capability. The Smart Display can be used as your PC's primary monitor when it is docked, but when you remove the LCD panel from its docking station, the PC switches its display from a local VGA connection to a remote desktop connection using the embedded 802.11b radio inside the monitor. Now you can move around the home or office but stay connected to your PC as long as you are in range.

In the enterprise environment, where wireless coverage is good, you can move anywhere in the facility and have access to the files and applications running on your desktop PC. All of these new form factors emphasize the need to be well connected internally in order to gain mobility. This means support for Wi-Fi is inevitable for consumer electronics devices. The reality today is that most CE companies have announced home entertainment products that will be enabled with Wi-Fi connectivity. Some already have it while others plan to include Wi-Fi as a standard feature.

In the home entertainment arena, Sony's new Linux-based digital video recorder (Cocoon) comes with a 160 GB hard drive and a high-speed Internet connection. It will likely sport a wireless interface soon. In addition, Philips' Streamium, a state-of-the-art audio microsystem that includes traditional microsystem features, such as an AM/FM tuner and a CD player, is available today. Streamium raises the stakes with the trio of Internet-radio reception, MP3 CD compatibility, and playback of any MP3s on networked PCs. It is also the first microsystem to offer full-blown DAR (Digital Audio Recording) functionality. All these devices assume a broadband high-speed connection through DSL or cable. Wi-Fi interfaces will be the next logical step for these devices. In addition, there is also another class of products that leverage existing home networks and help bridge or share content with PC desktops and other CE devices. Apple, HP, and Dell are reportedly working on such systems.

Wireless TVs have also hit the market in 2003. Sharp's AN-SS700, a Wireless TV system, is designed to operate in the 2.4-GHz band. The AN-SS700 includes a transmitter and receiver each measuring 5.83"(W) x 3.11" (H) x 1.54" (D). Specifications include the use of Sharp's MAC technology and an MPEG2 encoder/decoder for compressed audio and video. The transmitter connects to a VCR, DVD, or satellite tuner by a composite video signal. Digital video is wirelessly transmitted, and can be viewed remotely using the LCD TV connected to the receiver unit. A wireless remote control can be used to operate the transmitter. Finally, another class of entertainment devices is also becoming Wi-Fi enabled. These are the MP3 players, media nodes, gateways, and also gaming consoles.

The partnerships of traditional wireless LAN IC providers with consumer electronics companies is example of how the Wi-Fi-related industry vendors are attempting to broaden their customer base and create a market for non-PC-based applications of wireless LAN technology. This trend is expected to broaden into 2003 and beyond. Possibly the weirdest use of Wi-Fi yet is a new wireless sprinkler system that is coming out in mid 2003 from Digital Sun. The X.Sense creates a wireless mesh network using moisture sensors strategically inserted into your lawn that keeps track of when and how much to water. In case the owner is afraid of hackers breaking into the sprinkler system, the wireless network is encrypted.

6.5.3   The Mobile Handheld Market

The convergence of PDAs and cellphones, has further spurred the growth of converged phones and PDAs. Known by various names, these wireless LAN smart phones and PDAs will further extend the life of 802.11b solutions. The key differentiators from the computing market are overall system power, multimedia, compactness, and interface types. Already, today's PDAs and Pocket PCs come equipped with CompactFlash plug-ins and Wi-Fi attachments. The next-generation PDAs and Pocket PCs will likely embrace TI's OMAP and Intel's MSA.

From a wireless LAN IC perspective, there are two main classes of customers: the cellular handset manufacturers and the PDA customers. The cellular handset manufacturers include companies like Samsung, Nokia, Motorola, SonyEricsson, NEC, and Siemens. The PDA customers include HP, Dell, Sony, and Palm. Although this market appears nascent, the upside potential can be huge when 802.11 low-power chip sets are ready. The cellular handset market is by far the largest market compared to PDAs and notebooks. The key differentiators from the computing market are performance and features. The most challenging feature is power consumption. In order to minimize the host processing, most mobile handheld devices adopt a zero overhead host strategy. This typically translates to an ARM processor. Whether it is an ARM7 or an ARM9 depends on the processing requirements, licensing, and overall cost.

There is some differentiation when it comes to radio architectures as well but it is not a really big issue. The types of interfaces to the host are also important due to the compact nature of such devices. Typically, there are only a handful of operating systems that support such devices. They are Pocket PC, Symbian, Palm OS, and Linux. Due to the nature of applications that reside in converged phones and PDAs, it is highly conceivable that VoIP using 802.11 will be one of the key drivers for this market. Hence, there is a need to support QoS and security on these devices. In addition, the choice of form factor will be critical in order to support the existing device and also future-generation devices. Finally, the ability to coexist with cellular and/or Bluetooth will be prime considerations for design success as well. Figure 6.6 illustrates the mobile handheld projections.

6.6   High-Throughput Wireless LAN Market

The discussion of high-throughput wireless LANs a couple of years ago essentially drove the frenzy over 5-GHz technology, with silicon start-ups coming out of the woodwork in 2000 and 2001, and announcing plans to develop 802.11a solutions. Although networking companies perceive 802.11b technology as providing suitable bandwidth for everyday data transmission, consumer electronics companies were not satisfied with the 5 to 6 Mbit/s throughput that 802.11b generally provides. Then came 802.11g attempting to be the disruptive technology for the 802.11 series. Studies conducted thus far indicate that throughput degradation with range for 802.11a, g, and b does not put 802.11g in any better position. Range is a significant concern with 802.11a, especially for home environments, which often involve structurally diverse arrangements with a variety of materials through which radio transmissions must pass. Even though wireless LAN is making headways into the home and SOHO environment today, the primary applications have been confined to computing. Typical applications include e-mail, file sharing, and the occasional video conferencing (NetMeeting etc.). Whereas this represents the state of home networking applications today, the future looks radically different. The market after 2003 will be mostly driven by the increase of infotainment throughout the home. The key driver here will be the distribution of audiovisual information.

Thus far, most of the activities surrounding high-throughput wireless LANs have been confined to the standards arena, in particular the IEEE 802.11. This is quite typical for most technologies as the source of incubation starts with innovation in R&D labs worldwide. In the case of IEEE, they have created a Task Group, called the IEEE 802.11n (part of the IEEE nomenclature for naming Task groups; the previous Task group approved was IEEE 802.11m) in early 2003. This Task Group, aptly called the High Throughput Task Group, has been chartered to look into the technology requirements needed to create a new addendum or standard that will pave the way for a new generation of devices that will bring a new level of experience to wireless networking. As of the writing of this book, the group has just started with its activities.

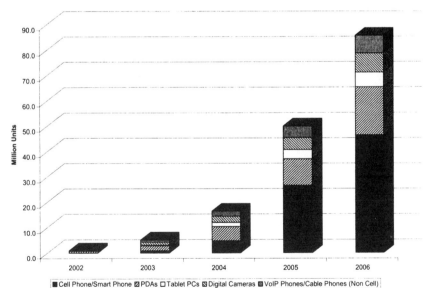

Figure 6.6: Projections for Mobile Market.

6.7   The Handheld Universe is Converging into a Single Device

Most NICs shipped to date have been in the form of PCMCIA cards, and this form factor will continue to dominate for the next three or four years. The mini PCI form factor is gaining traction and is expected to dominate in 2003. However, companies in the market of developing wireless LAN PDAs and converged cellphones are looking at other form factors and interfaces, such as Secure Digital Input/Output (SDIO) and Compact Flash (CF), for reasons of compactness. The SDIO form factor is the latest entrant into the 802.11 markets.

In late 2002, SyChip (a fabless semiconductor manufacturer) launched its WLAN6060EB and WLAN6060SD end-to-end, plug-and-play Wi-Fi solutions for embedded and SDIO-based applications, respectively. SyChip is also in the business of providing reference designs for a secure digital (SD) input/output NIC for OEM vendors that supports Serial Peripheral Interface (SPI), 1-bit, and 4-bit SD transfer modes. Sychip's modules start with Intersil's Prism 3 chip set and add the company's own software, packaging, and power-control expertise to realize what SyChip calls a complete solution. The WLAN6060EB module, measuring 25 x 23 x 3 mm, is designed to be embedded within mobile handsets, PDAs, and ultrathin notebook computers. It requires only an antenna to complete the signal chain from the media-access control outward. The design will also support PC Card and CF+ interfaces. The standby current is less than 70 milliamps (mA) and the transmitting current consumption is 287 mA. The wireless LAN 6060SD module has similar specifications but comes with a Secure Digital I/O interface and finished SDIO form factor. The wireless LAN 6060SD is gaining traction as several equipment vendors have announced 802.11b wireless LAN SDIO cards based on its design. In early 2003, manufacturers announced product launches based on the 6060SD including SanDisk and Socket Communications. SanDisk and Socket released their CF Wi-Fi cards at almost the exact same time in 2002. In 2003, both companies announced they will release SD Wi-Fi cards based on SyChip's WLAN6060SD design, for SDIO equipped handhelds. The key difference this time is that Pocket PCs and Linux PDAs are not the only target markets. In 2003, Palm's Tungsten T is a prime target, as are several Palm OS 4 devices. As market consolidating occurs, and more and more standards fall into place, companies like Sychip get to decrease their product range, while selling more units. Instead of an SDIO Wi-Fi card for Pocket PCs, and another for Palms, they can release one that works on both, because the market has standardized, somewhat, on the SDIO interface. This is happening more and more from choice in processors to screens to specs of all sorts. By the end of 2003, Sychip expects more than 10 OEMs to support WLAN6060SD. Designed for 802.11b applications, SyChip's WLAN6060SDIO reference card design is developed around the company's WLAN6060B module that includes a baseband processor, 802.11b MAC, memory, voltage-controlled oscillator, 802.11b transceiver, antenna switch, power amplifier, and software drivers. The reference design also includes SDIO Now! software developed by BSquare. Companies like Socket will use SyChip's reference during the development of cards for PocketPC applications.

6.8   The Digital Interface

In an effort to develop open industry interface specification standards between the MAC-layer and PHY-layer, as well as between the Baseband and Radio blocks for wireless LAN, the Jedec-61 Committee was formed in 2002. Essentially, its charter is to develop well-known interfaces between the various subsystems of a wireless LAN system. As shown in Figure 6.7 for the Superhet and Near Zero IF and Zero IF, the goal is to define a digital interface between the RF-Baseband and Baseband-MAC. This is needed because signaling requirements between the major functional blocks in wireless LAN solutions, especially between the Baseband and Radio, have unique requirements. A high-bandwidth, dedicated, point-to-point interface with low overhead and minimal gate count are key requirement for many of the intended applications. Due to the real-time nature of data communications between the Baseband and Radio, a low latency interface with no contention is essential.

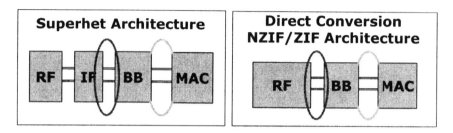

Figure 6.7:   Interfaces to be Defined for Wireless LAN Architecture.

If the JEDEC 61 efforts materialize, the current mini-PCI wireless LAN card will be replaced by several new potential industry standards now being drafted by the committee. The JEDEC Solid State Technology Association JC61 committee is expected to finish several draft standards later this year for new interfaces between the baseband, MAC, and RF portions of wireless LAN chips. The primary motivation for the new standards is to open the wireless LAN market to allow the mix-and-match of different vendor RF, baseband, and MAC chips for greater competition. In essence, the implications of the standard will allow moving the radio in PCs close to the system antenna to eliminate the relatively long coax cable now used to connect the mini-PCI card to the antenna in PCs. This will hopefully reduce thermal problems, RF interference, and power consumption resulting from the long coax cable needed to connect the miniPCI to the antenna. The first draft slated to be completed in Q303 will define a new interface between the baseband and RF. One architecture will move the baseband near the radio and antenna to get rid of the mini-PCI card coax cable.

Another draft standard expected to be completed in Q403 will define a new interface between the baseband and the MAC. A third draft will divide the MAC into two halves: upper MAC and lower MAC. The upper MAC for non-time-critical functions, such as encryption, and a lower MAC for time-critical operations, such as turning on and off the transmitter. The lower MAC can be sited with the baseband and RF close to the antenna, while the upper MAC can remain elsewhere in the PC box.

6.9   Summary

The insatiable demand for wireless connectivity will further spur the growth of wireless LANs. The expansion will occur in new markets such as the mobile handheld market, Wi-Fi enabled consumer electronic market, and next-generation computing market. The drive toward lower cost will lead to further integration of components, newer manufacturing processes and densities, and smaller form factors, especially for the handheld market.

References

[1] 3Q 2002 Wireless LAN Market Analysis, Report IN020202WL, In-Stat MDR.

[2] "Its Cheap and It Works: Wi-Fi Brings Wireles Networking to the Masses," December 2002, Report IN020181LN, In-Stat/MDR.

[3] 2003 Outlook on Wireless LAN Semiconductors, WR February 2003, Hambrecht & Co.

Chapter 7

Wireless LAN IC Industry

In spite of the global economic downturn, which has impacted the technology sector, one bright spot that has bucked the trend so far is the wireless LAN semiconductor market. Key factors fueling this growth in the wireless LAN IC market are the adoption of wireless LANs within the enterprise and home. On the carrier side, wireless LANs are being touted as a public hotspot complement to 3G cellular services. The back-end impact of this is increasing industry support from major OEMs and ODMs. In addition, an increase in the number of wireless LAN IC startups and a major shift in strategies from industry heavyweights have upped the ante for better-performance and lower-cost chipsets. A report from U.S. market research firm In-Stat/MDR shows that in 2002, wireless LAN chipset sales are forecast to top 14 million units, an increase of 75% from 2001. Part of that huge demand should come from Taiwan. This trend will be covered as part of the market forecast. We will also examine the overall trends in the IC market and look at the demand and supply side. In addition, we will examine the semiconductor components of a wireless LAN system. Lastly, this chapter will profile the wireless LAN IC players that will likely shape the future of this industry.

7.1   Market Forecast and Trends

Generally, this is the first question investment/business firms consider before entering the market. To appreciate the size of the market, one has to go back a little in time. Wireless LANs have existed since the early 1990s. Most users of wireless LANs back then were limited to very few specialized vertical applications like manufacturing. When the 802.11b standard was finally ratified in September 1999, it started a new era for the wireless industry as a whole. Much venture capital and angel funding were channeled toward the wireless LAN IC industry in 2000, the first full year that 802.11b chipsets were available. In 2000 alone, more than 8 million ICs were sold, up from just over 2 million the year before.

Figure 7.1 shows the growth potential for wireless LAN ICs globally by the three main sectors: computing, consumer electronics, and mobile handheld. The difficulty in collecting objective data is that the numbers from market research firms vary tremendously in spite of the similarities in methodologies. This difficulty is compounded by the fast changing wireless LAN landscape. For example, the wireless LAN IC industry in Taiwan is often overlooked in terms of market strength. According to the Taiwan Market Intelligence Center (MIC) estimates, Taiwan accounts for about 80% of the global market for wireless LAN ICs. In 2003, Taiwan is expected to have about 83% of the global market share. This should not be too surprising given the island's strength in wireless LAN card/system and the notebook industry. 2003 should be a breakout year for Taiwan's entry into the global wireless LAN 802.11b marketplace as further commoditization occurs in the Wi-Fi industry.

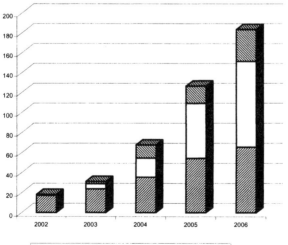

Figure 7.1:   **Wireless LAN IC Growth Potential for Computing, Mobile, and Consumer Markets.**

The rapid adoption of wireless LANs has spurred the demand side of wireless LAN ICs, in particular the client market. The infrastructure market is also poised to experience significant uptake but the volume is significantly smaller. The client market is expected to have a higher demand due to the "Wi-Fi-cation" of many consumer electronic devices such as camcorders, cameras, set-top boxes, MP3 players, DVD players, audio systems, and even TVs. In addition, the ratification of 802.11g and the emergence of dual-band products 802.11a/b and 802.11 a/g will further boost the demand side of the wireless LAN IC market.

In the chipset arena, the entry of new 802.11b providers like TI and Broadcom also helped push chipset prices further, adding price pressure on incumbents like Intersil and Agere. It is also clear that HiperLAN/2 and HomeRF will not be major wireless LAN standards based on market demand for such products. This is very reminiscent of the ATM and Ethernet wars that were fought several years ago. Again, simplicity prevailed. ATM, with its solid mechanisms, (albeit complex) for supporting real-time and isochronous traffic, was deemed to be far superior technically to Gigabit Ethernet, which offered virtually no QoS support then. The market today tells the full story. Ethernet won the enterprise war. In a similar vein, HiperLAN/2, with its different classes of service to support QoS, was technically superior to IEEE 802.11. The reality of the marketplace today is still favoring simplicity. In addition, there are provisions made to extend the 802.11 standards to support some of the requirements of the European Regulatory Committee. This is further explained in Chapter 2.

7.2　Dissecting the Wireless LAN System

It is important for the reader to have a basic appreciation of the general components that make up a wireless LAN system. At the most basic level, the three building blocks are the radio, Medium Access Control (MAC), and the baseband. The radio front-end is composed of three discrete components: power amplifier (PA), intermediate frequency (IF) converter or Zero IF (ZIF), and the modem. The PA's main role is in signal amplification. It amplifies the RF signal from the RF/IF converter before handing it over to the antenna. On the reverse flow, it amplifies the received signal before handing it over to the RF/IF converter.

Typical characteristics of PAs include the level of integration, which may feature a three-stage design: power amplifier with on-chip input and interstage match, output power detector, and all necessary power management circuitry. The power detector typically provides over 20 dB of dynamic range with ±0.8 dB accuracy, allowing easy implementation of the accurate automatic level control (ALC) function without additional components. The PA also features an external bias control pin. Through an external digital-to-analog converter (DAC), the current can be throttled back at lower output power levels while maintaining sufficient adjacent channel power rejection (ACPR) performance. As a result, maximum efficiency is maintained at all power levels. Newer PAs deliver higher gain while meeting IEEE 802.11b spectral mask requirements and also reasonable saturated power. The high-output performance also makes it ideal for use in the IEEE 802.11g 54 Mbit/s extension, where the OFDM signal requires greater back-off and higher linearity performance. This is especially critical for PAs used in 802.11g and 802.11a applications, where the OFDM signal peak-to-average ratio for a 1% complementary cumulative distribution function (CCDF) is around 7.4 dB. Another aspect of PA design affecting performance is power consumption, which indicates how much current is drawn at a particular voltage in order to produce a specific output power. Output power figures typically range from +17 to +20 dBm. The design process for PAs also plays a role in determining the final BOM. The design process used can be silicon—germanium (SiGe), bipolar CMOS (BiCMOS), indium—gallium—phosphide (InGap), heterojunction bipolar transistor (HBT), and complementary metal oxide semiconductor (CMOS). HBT is typically used for optimal efficiency/linearity tradeoffs under stringent thermal conditions.

The next component is the radio. This usually involves a choice of two architectures: superheterodyne or ZIF. The superheterodyne (or dual-conversion) transceiver is considered the classic radio architecture, in which the received signal is down-converted to baseband frequency in two stages. Basically, a frequency is generated by a synthesizer and combined with the incoming RF signal to produce what is known as an IF. This is achieved by a two-stage receiver and transmitter architecture using an RF block to convert the incoming signal to an IF, in which image suppression and channel selection are performed with a narrow channel-select filter, such as surface acoustic wave (SAW) or ceramic filters. This stage is required because the original 2.4-GHz frequency received by the antenna is too high to be acted upon by the next stage. The now-filtered signal is then further down-converted to the baseband frequency, which is then digitized and demodulated in a DSP. This radio architecture has been used for decades, in part due to its excellent sensitivity and selectivity characteristics. This, however, comes at the expense of more complexity and cost, for such radio implementation typically requires an RF chip and an IF chip as well as discrete SAW filters and voltage-controlled oscillator (VCO)/synthesizers.

The ZIF radio front end is becoming increasingly popular due to lower costs and fewer discrete components. The ZIF radio transceiver has a direct-conversion architecture, meaning that it utilizes one mixer stage to convert the desired signal directly to and from the baseband without any IF stages and without the need for external SAW filters. Most ZIF radio designs also integrate the low-noise amplifier, VCO, and the baseband filters on a monolithic die. In fact, such integrated single chip ZIF transceivers have been proven for many years in cellular and pager applications and they are beginning to emerge in wireless LAN radio designs. ZIF radios require less external components and therefore, tend to reduce cost and size for the overall system. There are however, design challenges for ZIF architectures. For example, dc offset, flicker noise and local oscillator (LO) pulling. Dc offsets are mainly generated by the LO leakage, which self-mixes, thereby creating a dc component in the signal chain that affects the receiver performance and can cause the RF stages to saturate. Flicker noise, also known as 1/f noise, is low-frequency device noise that can corrupt signals in the receiver chain. Flicker noise is more pronounced with the ZIF architecture because of direct conversion to the low-frequency baseband. Another concern with direct conversion is the pulling of the LO by the PA output, which affects the direct up-conversion process. This is because the high-power PA output, which has a spectrum centered around the LO frequency, can disturb ("pull") the transmitter VCO. Recent advances in radio and modem designs are able to resolve these matters through a combination of proprietary radio design techniques and system algorithms in the baseband. For instance, dc offset can be addressed via a compensation scheme in which the offset is measured and reduced via unique radio and baseband algorithms.

The bottom line in these design trade-offs is cost versus complexity. The BOM for a dual-conversion radio design is generally more expensive than for a direct-conversion design in a single-band wireless LAN interface card, but the cost discrepancy is increased when the chipset handles two bands. A dual-band design using a superheterodyne transceiver needs a 5-GHz RF stage and a 2.4-GHz RF stage, discrete IF synthesizers/VCO, two SAW filters for image rejection and for channel selection for each band, and a common IF block driven by a discrete IF VCO. Assuming that this architecture uses an integrated modem/MAC IC, a dual-band chipset solution will require nine components. Compare this with the ZIF architecture, in which the lowest-cost dual-band approach is to combine the two ZIF transceivers into a single dual-band radio built on CMOS. This is possible because of the direct conversion feature of the ZIF transceiver architecture and the integration capability of CMOS process. In fact, the 2.4- and 5-GHz ZIF transceiver circuitry can be laid side by side on a monolithic die without much impact on the die size or package cost. This approach results in a two-chip solution without the need for external components such as SAW filters.

In summary, we have two architecture options. The first choice is the superheterodyne architecture, which spreads the channel filtering among multiple stages. The second choice is the zero or near-zero IF architecture. The trade-off here is between more components that require less linearity in the superheterodyne case while providing better performance. The ZIF option allows for fewer components but in turn has very stringent linearity and filter cutoff characteristics. The near-zero IF option performs worse but has the benefit of lower component count.

The last component of the radio front-end is the modem, which takes the received analog radio signal and converts it into a digital signal before handing it over to the baseband processor. The reverse flow involves taking the digital signal from the baseband processor and modulating it into an analog signal before handing it over to the IF converter.

7.2.1 General Design Considerations

The core functions of wireless LAN devices are embodied by the semiconductor inside them. Hence, knowing where wireless LAN products are headed can provide pretty good insights into where the semiconductor market is headed. In designing good wireless LAN chipsets, there are some general guidelines that most wireless LAN IC vendors follow. Although this section is not designed to provide an exhaustive treatment for the detailed design of wireless LAN ICs, it will provide the reader with a good understanding of the complexities and design trade-offs that exist today. Wireless LAN ICs define the core function of the network infrastructure and client in a wireless LAN environment. Even though wireless LAN IC designers continue to innovate to meet the increasing demands for enterprise, home, and mobile handheld applications, there is still room for further innovation. An example of this is the coexistence of wireless LANs with Bluetooth and 3G cellular technologies.

From a process perspective, many new entrants that have announced chipset availability for 802.11a/b/g are using low-cost, readily available digital CMOS process technology to implement highly integrated chipsets. Even though CMOS has the potential of reducing product costs, there is the risk of manufacturing yields offsetting prewafer costs and poor performance results. Typically, the CMOS process is optimized for maximizing digital switching performance as opposed to analog functions, hence the consistency and performance of analog functions are worse than if the functions were implemented in an analog-based technology. CMOS-based radios and PAs will not necessarily reduce cost at the same rate as digital components produced using the same process.

There is also a growing trend toward "module technology," which essentially means a plug-and-play component consisting of all necessary parts to perform the necessary function. The most common approach so far for module technology has been on the RF side of the solution. An RF module is a plug-and-play component that by itself realizes a critical RF function. Development of an RF module requires specialized knowledge and experience to implement, test, and manufacture. By plug-and-play, it is meant that no tuning or adjustments are required and no external components are needed. The critical RF function realized means that the module has significant added value on time-to-market, die area, component count, BOM, test cost, and yields. In the baseband, advanced CMOS integrated circuit processes have shrunk the silicon real estate required for the processor, memory, and interface ICs. In the radio transceiver space, the move from a super-heterodyne to a near-zero IF radio architecture eliminates the set of passive components required for the IF functions. This trend has been accelerated by the steady improvement in the performance of the passive components integrated in the RF IC processes, thereby enabling the integration of passive components on chips where previously only external SMD components could meet the required RF performance level.

7.2.2   Putting the Pieces Together

We will illustrate how the baseband, RF, and PA work together to produce a full system solution. At the time of this writing, the only available product for a dual-band solution is from Atheros Communications, which we will use as an example. We chose the miniPCI Type IIIb form factor for illustrative purposes. This type of card provides the functionality of 802.11a/b to the host PC. The MiniPCI Specification 1.0 defines an alternate implementation for small form factor PCI cards. This specification uses a qualified subset of the same signal protocol, electrical definitions, and configuration definitions as the PCI Local Bus Specification 2.3. In the 5-GHz band, the 802.11a/b

MiniPCI Card uses all 12 IEEE 802.11a channels, but does not support the high-power modes in the upper channels. The 802.11a/b MiniPCI Card also supports tuning to the Japan-allocated 5-GHz frequency band of 5.17 to 5.23 GHz. The 802.11a/b MiniPCI Card uses the MiniPCI socket on a Microsoft Windows-based host computer to provide wireless networking to the computer. Once the Network Driver Interface Specification (NDIS) driver is installed and configured, the 802.11a/b MiniPCI Card allows peer-to-peer (ad-hoc mode) connections with other computers with 802.11a or 802.11b products installed, and also allows connections between the host computer and 802.11a or 802.11b-based access points (infrastructure mode).

The IEEE 802.11a/b MiniPCI Card operates from the PC host power supply and comprises a baseband section and an RF section. The baseband section deals with the interface to the host PC via the MiniPCI interface, and provides the data formatting, encoding, and encryption required by both IEEE 802.11a and IEEE 802.11b standards. Selection between IEEE 802.11a and b is determined within the NDIS driver configuration on the host PC using information provided by the baseband processor of the 802.11a/b MiniPCI Card. An Electrically Erasable PROM (EEPROM) memory device holds configuration data including the MAC address of the 802.11a/b MiniPCI Card.

The baseband processor is the origin and destination for all the front-end signals. Both transmit and receive signals are switched and transferred either to a 5-GHz front-end or to a second chip (2.4-GHz RF transceiver), which up-converts or down converts the 5-GHz signals to 2.4-GHz. The 2.4-GHz RF transceiver, in turn, feeds a 2.4-GHz front-end. Both the 2.4-GHz and 5-GHz front-ends are combined via two diplexers, which are designed to feed two dual-band diversity antennas. The 5-GHz signal is equivalent to the IEEE 802.11a signal, whereas the 2.4-GHz signal is equivalent to the IEEE 802.11b signal. The 5-GHz transmit signal is filtered after passing through an RF switch. The filter removes the different by-products of the internal 5-GHz RF transceiver LO. Also, the 5-GHz transmit signal is boosted with a PA. The PA drives a coupler/detector assembly. The coupler/detector's function is to sample the transmit signal and rectify it. The rectified signal is proportional to the output power and is used for power leveling and control. The 5-GHz transmit signal then passes through a bridge switch. The bridge switch is a diversity (transfer) type and has 4 ports: two inputs (transmit and receive) and two outputs (antennas). It enables the connection of any of the input ports to either one of the outputs. The 5-GHz transmit signal is then transferred through a low-pass filter (LPF), which removes any 5-GHz harmonics generated by the PA and bridge switch. The 5-GHz transmit signal then passes through a diplexer to the antenna port. The diplexer has a common (antenna) port and two more ports for 5 GHz and 2.4 GHz. The diplexer is transparent to 5-GHz signals between the common and 5-GHz port. Likewise, the diplexer is transparent to 2.4-GHz signals between the common port and the 2.5-GHz port. The diplexer includes a low-pass filter for 2-GHz harmonic rejection. The 5-GHz received signal is transferred in a reverse order from the antenna through the diplexers and bridge switches. It is filtered to reject image frequencies via a band-pass filter (BPF) and boosted via a low-noise amplifier (LNA). The transmitted and received 2.4-GHz signals follow a path similar to the 5-GHz signals in the 2.4-GHz front-end.

7.3   Vendor Profiles

The wireless LAN IC market is entering the first stages of a shakeout, but that is not stopping several major semiconductor suppliers from joining one of the electronics industry's most promising growth areas. Some of the notable acquisitions in 2002 include RFMD's acquisition of Resonext and Philips' acquisition of Systemonic. These fundamental shifts in the wireless LAN landscape presents opportunities that can be highly rewarding provided timely execution of products can be maintained. Even chip giant Intel has announced big plans to penetrate the market. AMD has also made similar announcements to produce wireless LAN chipsets. At the same time, major existing players, including Broadcom Corp., Intersil Corp., and Texas Instruments Inc., are all in the process of introducing new versions of their wireless LAN offerings as the pressure to implement multiband solutions intensifies. In October 2002, RF Micro Devices agreed to acquire Resonext Communications Inc., which had just introduced a dual-band wireless LAN solution. Agere Systems, which had been an early participant in supplying 802.11-compliant devices, formed an alliance with Infineon Technologies A.G., Munich, Germany, to coproduce wireless LAN chipsets. Intel Corp., which has yet to ship an internally developed wireless LAN device, has made the technology a cornerstone of its communications and mobile computing effort. The company announced it will invest $150 million over the next few years in companies developing wireless LAN-related technologies. To date, the wireless LAN market has attracted as many as 30 chip vendors, but the tally board changes monthly as companies acquire competitors and enter and exit the market.

7.4   Advanced Micro Devices (AMD)

AMD is a leading supplier of microprocessors and flash memory devices for computing and communications applications. Headquartered in Sunnyvale, California the company has manufacturing facilities in the U.S., Europe, Japan, and Asia. AMD's Alchemy division has an offering for Wi-Fi in PDAs as well as notebooks and other devices.

7.4.1   Company Overview and Strategy

AMD is pursuing a strategy similar to Intel's in this lucrative Wi-Fi chip market. Not to outdone, AMD also has Wi-Fi products. Its current offerings include a reference design kit for a miniPCI card. This mirrors the Calexico miniPCI card that was announced for Intel at about the same time. The reference design is for a Wi-Fi adapter that takes only one side of a 2.8 by 1.8-inch miniPCI card.

The AMD Alchemy™ Solutions Am1772™ wireless LAN chipset and a miniPCI card reference design kit is currently being sampled and is expected to be production ready by Q103. The key attributes of this chipset are lower system cost, longer battery life, and lower host CPU utilization. The Am1772™ wireless LAN chipset is designed to enable customers to make easy transitions to future 802.11 MAC enhancements. It is a CMOS solution that utilizes a baseband processor and a MAC, with a descriptor-based direct memory access (DMA) host interface and on-chip hardware acceleration to reduce the host CPU load. The CMOS design helps reduce power consumption for the miniPCI-based reference design to 134 mA when receiving and 232 mA when transmitting. This represents an improvement over competing designs, resulting in increased battery life for mobile applications. In addition, the highly integrated solution reduces the number of additional components and enables a very compact design.

The Am1772 chipset comprises the AMD Alchemy Solutions Am 1770™ RF transceiver and the AMD Alchemy Solutions Am1771™ baseband processor and MAC. The Am1770 transceiver utilizes direct down-conversion, which eliminates the requirement for an intermediate frequency (IF) chip. It also has an integrated loop and based band filters with auto calibration. In addition to the programmable LAN, it also has a programmable low-power CMOS design. Another feature that stands out for the Am1772 is the digital interface to the Am1771 BB/MAC. This feature helps eliminate the need for an on-chip microcontroller and external flash memory and SRAM. Finally, it also has an on-board AGC and requires no baseband intervention. The Am 1770 comes in a 7 mm x 7 mm dimension in a 48-pin low-profile quad flat pack (LQFP) package. The Am1771 chipset is an integrated baseband/MAC that features on-chip hardware acceleration designed to significantly reduce host CPU load. The use of the baseband processor and MAC with a descriptor-based DMA architecture also benefits the customer by enabling lower system costs through the elimination of the on-chip microcontroller and the associated nonvolatile (flash) memory and SRAM. The use of auto-calibration technology also reduces the need for costly and time-consuming system calibration during the manu-facturing process. The Am1771 baseband/MAC comes in a 13 mm x 13 mm 176 low-profile, fine-pitch ball grid array (LFBGA) package.

7.4.2   Product Overview

Widespread availability for sampling of the Am1772 chipset and Reference Design Kit (RDK) is scheduled for early 2003 and production availability is planned in Q1 2003. AMD says the chipset will let OEMs lower device costs by using fewer components, as well as extending battery life through lower power consumption. The target customers are both embedded manufacturing partners, such as Ambit and Z-Com. By the second half of 2003, AMD hints, not only will it add an 802.11a+b/g chipset to the Alchemy Solutions line, but 802.11b will have joined USB, Ethernet, LCD and memory controllers, and other capabilities in its system-on-a-chip (SOC) MIPS processors such as the current Alchemy Au1100.

7.5   Atheros Communications

Atheros Communications was founded in May 1998 by some of the leading experts in RF and signal processing from Stanford University, the University of California at Berkeley, and private industry, including several former 3Com employees. In the 5-GHz and dual-band markets, Atheros is the incumbent because companies such as Sony, Intel, Proxim Netgear, and SMC Networks are shipping wireless LAN gear based on the Atheros 802.11a chipset. As of early 2002, Atheros was still the only company shipping 802.11a chips. In addition, Atheros also scored another major victory when it announced the availability of the first tri-mode solution supporting 802.11a, 802.11b, and pre-802.11g extensions in March 2002. In 2001 alone, Atheros posted sales of $4.3 million in 802.11a sales. Atheros' first mover advantage in the wireless LAN market has led to an impressive list of partnerships with many of the world's leading wireless equipment manufacturers including Accton, Actiontec, Acrowave Systems, AirVast Technology, ALPS, Ambit, Askey, ASUS, BenQ, Contec, CyberTAN, Delta Networks, D-Link, Gemtek, Gigabyte Technology, Global Sun Tech, HP, IBM, ICOM, Intel, Intermec, I-O Data, Linksys, NEC, NETGEAR, Philips, Proxim, Samsung, Senao, SMC

Networks, Sony, Symbol, TDK, TECOM, USI, W2 Networks, Wistron NeWeb, Z-Com, and others.

7.5.1   Company Overview and Strategy

In a little over four years, Atheros has established itself as the wireless LAN market leader in 802.11a and 802.11a/b/g. The journey to the top was not exactly an easy ride, though. The engineering community had criticized its first 802.11a chipset. Many engineers claimed that an all-CMOS design will not produce optimum performance at 5-GHz, and several felt that an integrated CMOS power amplifier wasn't the best choice. As a result, the chipset currently also supports an external PA. In spite of these initial engineering setbacks Atheros garnered an impressive list of design wins from D-Link, Intel, Proxim, Sony, Netgear, TDK, and others. Atheros' latest product, the AR5001 chipset, should only help the company further strengthen its position in the market. Its second-generation product should help ease fears of those manufacturers who hesitate to go with a first-generation product, and its 802.11a+b+g combo chipset, the first of its kind, will be welcomed by those manufacturers that "want it all" and that are looking for a way to support 802.11a while still supporting 802.11b legacy access points.

Part of the reason Atheros has been able to keep its leadership status is as follows:

- Leadership: first mover, trendsetter in 802.11a and 802.11 a/b/g
- Excellent relationship with media
- Strong promoter of interoperability activities (e.g., WFA testing)
- Captured all major module manufacturers
- Strong emphasis on OEM partnerships
- Excellent customer support and relationship management
- Most design wins in the industry for 802.11a modules
- An impressive list of partners/customers

Even though Atheros enjoys the leadership position today, competitors (both startups and established) are hot on their heels today. Atheros has a first mover advantage, but Intersil has deeper relationships with many of the biggest makers of wireless networking equipment.

7.5.2   Product Overview

Atheros' product line can be classified into two categories: 802.11a and 802.11a/b/g. The 802.11a product is in its second generation now. It features the AR5001A chipset, and similar to the first generation (AR5000) before it, uses an all-CMOS design and is composed of two chips, an RF transceiver and a baseband/MAC. Unlike the AR5000, the AR5001A also supports the upper 5-GHz U-NII band, as well as the Dedicated Short-Range Communication (DSRC) band at 5.9 GHz. This 75-MHz band has been allocated for automotive communication applications, such as automated toll collection, navigation downloads, and even multimedia downloads. The chip also supports an Atheros proprietary 108 Mbit/s Turbo Mode. In addition, Atheros also produced an Access Point version of their chip (AR5001AP), which adds required Ethernet functionality in the form of two Ethernet MACs.

The second category is the tri-mode solution. The AR5001X chipset not only supports 802.11a, but also 802.11b and pre-802.11g as well. The implementation uses a three-chip set, composed of a 2.4-GHz radio, a 5-GHz radio, and a baseband/MAC. The chipset cannot support 2.4-GHz and 5-GHz simultaneously, but this is generally not a requirement for the NIC applications that this chipset is designed for. The AR5001X chipset includes the AR5111, AR5211, and AR2111 components.

The AR5111 is the 5-GHz Radio-on-a-Chip (RoC) and has the following features:

- Dynamic IF Dual Conversion architecture provides super-heterodyne performance at ZIF prices
- Support for IEEE 802.11a standard
- Integrated second-generation PA and LNA
- External PA and/or LNA can be used for special applications
- Enhancements to the transmit and receive chains
- Eliminates many RF and IF filters; no external voltage-controlled oscillators (VCOs) or SAW filters needed

The AR5211 Multiprotocol MAC/baseband processor has the following features:

- Supports both 5-GHz and 2.4-GHz RoCs
- Smart Select™ technology automatically chooses the data rate, error-correction mode, radio channel, power-management method, and security technology best suited to any situation
- PCI 2.2 and PC Card 7.1 host interfaces with DMA support

The AR2111 2.4-GHz Radio-on-a-Chip (RoC) has the following features:

- Support for 802.11b and pre-802.11g extensions
- Advanced wideband Receiver with Best Path Sequencer for better range and multipath resistance than conventional equalizer-based designs

7.6   Intersil Corp.

Intersil's Wireless Networking Business unit is dedicated to the design and development of silicon technology for wireless LANs. Its flagship product, the PRISM® wireless LAN chipset, is widely recognized as the world's leading silicon solution for wireless networking systems and has been adopted by more than 50 of the world's leading telecom, networking, and computer companies in more than 100 product designs. Its market share for 2002 alone was about 60%, generating revenue of about $239 million.

7.6.1   Company Overview and Strategy

As a leading company in communications technology, Intersil's vision is driven by a three-prong strategy: high-performance analog, wireless networking, and creating synergy between analog business and wireless networking. Consistent growth, proprietary products, and a high-margin business fuel the performance analog business. Intersil's wireless networking strategy is built upon maintaining their leadership position in 802.11 (including 802.11a/b/g) and also providing the complete solution set (radio, PA, and baseband/MAC). Being able to cross-sell between these two markets has enabled Intersil to generate a pretty strong cash flow so far. As the incumbent leader in wireless LAN IC, Intersil's strategy execution so far has been pivotal to the company's success. Intersil invested in the technology as early as 1994 and was an active participant in defining the early architectures of 802.11. Intersil's strength lies in its ability to provide a complete solution. This capability was realized when it obtained MAC technology through the acquisition of Choice and No Wires Needed. These acquisitions allowed Intersil to provide complete 802.11b-compliant chipsets. Intersil scored well on the customer front as well. They have provided good reference designs backed by good support engineering staff. Lastly, Intersil is also aggressively pursuing the path of next-generation wireless LAN technology, always attempting to stay two steps ahead of the curve. Intersil Corp. laid claim to the first two-chip, dual-band wireless LAN chipset for 802.11a, b and g in Oct 2002. Intersil claimed its performance surpasses that of all other 802.11a, b, and g and dual-band wireless LAN devices. The company debuted the Prism Duette chipset at Comdex in November 2002, using it to wirelessly stream HDTV video at 2.4 GHz and 5 GHz.

In terms of market leadership, Intersil remained the wireless LAN IC leader in 2002. Intersil still maintains an impressive customer list, with their top customers for 2001/2002 as follows: Cisco, 3Com, Symbol, Samsung, Philips Components, and Nvidia in the USA. They have good customer relationships with Accton/Eumitcom and Golden Sun Technologies in Taiwan and Nokia in Europe.

In spite of aggressive competition, Intersil is still the dominant market player in the wireless LAN market. This is due to the size of the 802.11b market, which still accounts for the majority of 802.11 chipsets today. In addition, they have a very good marketing and sales organization. Intersil is a long-term player in the market, having invested in this market since 1995. They still maintain excellent partnerships with key customers and are also actively involved in driving standards committees and industry alliances such as the IEEE 802.11 and the WFA. Their future remains rosy in spite of the rapid commoditization of wireless LAN products and the steep challenge from Taiwan Inc. They have the opportunity to be the first to market with a two-chip complete 802.11a/b/g product. Intersil also has the wherewithal to further integrate the dual-band PAs into the RF front end for a more integrated dual-band solution at a price competitive point. The threat that will continue to challenge Intersil is their access to state-of-the-art fabs, which remains crucial if they wish to secure their long-term position as market leader. In addition, the market entry of Broadcom, TI, and Intel will further threaten Intersil's planned dominance of the dual-band market. To summarize, the challenge Intersil faces comes both the low-end NIC and the high-end access point market.

7.6.2   Product Overview

The PRISM Duette™ is a dual-band 5-GHz (802.11a) and 2.4-GHz (802.11b and 802.11g) wireless LAN solution The PRISM Duette features Intersil's BBP/MAC and direct down-conversion ZIF architecture. Intersil components used on the PRISM Duette radio reference design include:

- ISL3890: ARM9-based MAC/BBP supporting OFDM and CCK
- ISL3690: Dual-band ZIF
- ISL3990: Dual-band PA
- ISL3090: VCO

7.7   RF Micro Devices

RF Microdevices (RFMD) acquired Resonext Systems for $133 million in stock. This strategic acquisition provides RFMD with an immediate entry point into the lucrative wireless LAN multimode market. RFMD's core expertise is in the 802.11b radio, whereas Resonext's value proposition is the complete baseband, MAC, and RF solution. Prior to the acquisition, Resonext announced its combo solution in October 2002. This two-chip set supports both 802.11a and 802.11b simultaneously, and will also support 802.11g. This new RN5220 chipset uses the same basic baseband/MAC processor as the RN5200, but with the dual-band CMOS radio on a single chip.

7.7.1 Company Overview and Strategy

RFMD's expertise lies in RF CMOS, baseband/DSP, MAC, and protocol stack software. Their chipsets are integrated into wireless clients (PC cards and miniPCI cards), as well as specialty modules in consumer electronic devices, access points, and commercial and residential gateways. RFMD will expand its product line with wireless LAN chipsets that support multiple bands and multiple standards in the 5-GHz and 2.4-GHz bands. The future chipsets will be built upon the foundation of the RN5200 chipset, leveraging the Zero-IF CMOS radio, AccuChannel, Flexible MAC, and APoC architectures.

7.7.2 Product Overview

RFMD's offering of a dual-band solution is the RN5220 dual-band chipset. The chipset consists of a single-chip, dual-band CMOS radio and a multiprotocol baseband/MAC chip, both manufactured with the 0.18-micron, 1.8-volt CMOS process. Based on the company's 802.11a product, the RN5220 chipset is compliant with IEEE 802.11a/b standards, and is designed to fully support the 802.11g specifications with a data transfer rate of up to 54 Mbit/s. The chipset radio is based on zero-IF architecture and is hardware, MAC firmware, and driver compatible with the company's first-generation 802.11a chipsets, and maintains predominately the same feature set and interfaces. Its design supports a reduced bill of materials cost and low power consumption. The chipset features AccuChannel equalization technology, which increases the SNR by up to 4 dB and provides up to a 32 percent greater range for all data rates when compared to standard OFDM modems. A flexible MAC architecture supports customer differentiation and evolving IEEE drafts for security (802.11i), quality of service (802.11e), dynamic frequency selection and transmit power control (802.11h). The highly integrated, single-chip dual-band transceiver, built with the 1.8 V DC, 0.18-micron CMOS process, reduces system design complexity, lowers BOM cost, and features low power consumption figures of 200 mW (transmit/receive). The transceiver incorporates true zero-IF CMOS architecture, which directly converts 5-GHz and 2.4-GHz to baseband frequencies without any intermediate frequency stages and without the need for external filters. The transceiver also is designed to support the 2.4-GHz OFDM requirements for the 802.11g extension. The zero-IF architecture integrates the LNA, filters, PLL, synthesizers, and VCO required for dual-band operation with only a nominal increase in circuitry. This level of integration results in a single-chip dual-band radio with minimum external components.

The baseband/MAC chip is designed to support 802.11b at 11 Mbit/s, as well as 802.11a/g at 54 Mbit/s. The modem contains proprietary algorithms optimized for true zero-IF system performance with dc offset removal and IQ imbalance correction algorithms. AccuChannel equalization technology extends the range and data rates for wireless LAN deployment by counteracting the effects of signal degradation caused by real-world impairments such as multipath delays and signal attenuation. The RN5220 implements a fully programmable IEEE 802.11-compliant MAC running on an on-chip ARM processor. The MAC firmware can be provisioned for customer-specific differentiating features such as network performance enhancement, network management, and power control. The RN5220 chipset is made up of a two-chip combination of a transceiver and a baseband/MAC chip. The single-chip transceiver has a tuning range of 4.9 GHz to 5.4 GHz, consumes less than 200 mW in typical transmit and receive mode, and has a measured 7 dB noise figure for 802.11a and 6 dB for 802.11b.

7.8   Texas Instruments

Texas Instrument's foray into the wireless LAN market did not really take off until they acquired Alantro in June 2000. TI offers three high-performance 802.11-based wireless LAN solutions, the TNETW1130 (802.11a/b/g), the TNETW1100B (802.11b), and the ACX100 (802.11b).

7.8.1 Company Overview and Strategy

TI stepped into the Wi-Fi business in June 2000 and is considered a late entry for the DSP giant. TI now ranks fifth in Wi-Fi chip market share, trailing Intersil, Agere Systems, Atheros, and Philips Semiconductors. On September 3, 2002, TI introduced an integrated BB+MAC wireless LAN solution, TNETW1100B. The solution features Extra Low Power technology, which is important in battery life considerations for portable devices. In May 2002, TI announced a partnership with D-Link regarding integration of TI's ACX100 chip into 2.4-GHz D-LinkAir wireless network solutions. D-Link's solutions are targeted at the consumer market. Recent moves by Nokia to sell Wi-Fi phones and by cell phone service providers such as T-Mobile to add access to Wi-Fi networks has likely helped push TI into redesigning its chips for the mobile handheld market. The two key attributes TI touts in its new-generation chipsets are compactness and low power consumption. The TNETW series chips are about 44 percent smaller than the company's previous Wi-Fi chips (ACX), which were for laptop computers. That makes them a more comfortable fit inside the limited space of a cellphone or PDA. In addition, the new TI chips reduce power consumption tenfold. They go into a low-power "stand-by" mode when the device they're in is not logged onto a network, which on average is about 95 percent of the time.

Of all the companies that have announced low-power offerings for 802.11, TI is the one with the most aggressive approach. In September 2002, TI announced that it was sampling a revised version of its ACX100 wireless LAN chip, which consumes only a tenth as much power in standby mode as its predecessor. The combined baseband/media-access control chip supports 11- to 22-Mbit/second wireless connectivity and is aimed at such embedded portable applications as laptops, PDAs, and smart phones. TI sees Wi-Fi connectivity going well beyond laptops and into embedded applications. As a result, TI decided to respin the ACX100 baseband/MAC chip into a low-power version that can help them address both the embedded and convergence applications. TI devised a method to reduce standby power in the TNETW1100B. Basically, the reduction comes from carefully skipping network beacons. They also implemented a clocking mechanism that awakens the chip from sleep mode to handle transmission traffic and retain the chip's association with the network, then fall back to sleep. The design minimizes the power consumed in these periods through a rapid on-off cycling that wakes the chip up and puts it back into standby mode quickly. The design also features a hardware power-management scheme that shuts down certain blocks of the chip that are not used in standby mode. The end result is a chip that consumes less than 2 milliwatts at the chip level, or one-tenth as much as the ACX100. TI claims that the lower power consumption will support the addition of 802.11b connectivity to a PDA that will decrease battery life by only 2.1 percent, versus 26.8 percent with competitive solutions. The same connectivity will decrease battery life 1.9 percent versus 6.9 percent in competing solutions in laptops

7.8.2   Product Overview

TI's offering for the wireless LAN IC market comes in three flavors, each addressing a different market segment. The latest chipset, TNETW1130, is a dual-band offering featuring support for CCK, Barker, OFDM, and PBCC modulation schemes. The second chipset, the TNET1100B, is designed for the low-power and compact market. This is an increasingly important market for the mobile and consumer electronics space. Finally, the ACX101 is the first wireless LAN IC and was designed for WI-FI certification and also for cost-optimized reference designs for miniPCI, PC card, and USB applications.

TNETW1130 Solution
TI's TNETW1130 solution is a converged, single-chip MAC and baseband processor (BBP) for IEEE 802.11a/b/g. The device supports operation in the 2.4-GHz and the 5.2-GHz bands. Like other vendors who differentiate themselves by the ability to switch seamlessly between the two bands, TI's approach lies in its Auto-Band™ technology. Using Auto-Band™, the TNETW1130-based access points can simultaneously operate, and stations can automatically switch between the different transmission modes and

frequency bands, supporting 802.11b, 802.11g, or 802.11a. The TNETW1130 will support the four modulation schemes, namely OFDM, PBCC, CCK, and Barker in order to provide the dual-mode capability.

TNETW1100B Solution

TI's TNETW1100B is an embedded single-chip MAC and baseband processor that combines the low power consumption, small size, and superior transfer rates and range that are needed for embedded wireless LAN applications and consumer electronic products. The TNETW1100B is delivered in a 12 x 12-mm package that provides a 44 percent space savings over the 16 x 16-mm package used for the ACX100, though a pin-compatible version of the TNETW1100B in a 16 x 16-mm package is available as a drop-in replacement for the ACX100.

TI claims that using the TNETW1100B will allow for highly power efficient 802.11b connectivity with up to a 10-fold reduction in standby mode power consumption than competitive 802.11b chipsets for laptop computers, personal digital assistants (PDAs), cellphones and other portable devices. The lower power consumption allows end-users to be more mobile and productive longer with Wi-Fi equipment at home, work, or public hot spots. In addition, the host and I/O interfaces needed for both 802.11 station and access point applications support the following interfaces: 32-bit Cardbus, PCI, USB 1.1, CompactFLASH Plus, PCMCIA, and 16-bit generic slave.

ACX100 Solution

The ACX100 is a robust, reliable solution, offered with complete cost-optimized reference designs for PC card, miniPCI, and USB applications. Leveraging the architecture of TI's existing TI baseband processor, the ACX101, the ACX100 is a complete, single-chip MAC/PHY solution that combines high performance with robust functionality. It consists of an integrated MAC and baseband processor, combined with on-chip PCI, CardBus, and USB 1.1 interfaces, thus eliminating the need for external bridging components. The ACX100 is fully compatible with the current Wi-Fi-installed base.

7.9   Intel Corp.

As the 800-pound gorilla chip company, Intel's foray into the marketplace marks a significant shift in the chip giant's strategy for attacking the wireless LAN market. Intel has committed US$ 300 million to a marketing campaign to aggressively promote wireless LAN and its offerings. In January 2003, Intel announced it was branding its new wireless mobile computing initiative as Centrino. Centrino includes the Pentium-M (formerly Banias) CPU, the Calexico 802.11x Wi-Fi kits and notebook chipsets, the Calexico 802.11b and Calexico dual-band wireless kits, the PRO/Wireless 2100, and the PRO/Wireless 2100A. One of the early setbacks for Centrino was the bad press received on its availability. Wireless intelligence in Taiwan claims that Centrino (based on Intel sources) is slated to appear in March 2003. However, that schedule has been pushed back. Taiwanese wireless LAN suppliers are frantically bidding to supply modules, including Asustek, D-Link, and Gemtek Technology. In terms of partnering with Taiwanese manufacturers, Intel is clearly behind Broadcom and Intersil and will need to quickly ramp up to catch up. There are two wireless kits slated for the Centrino technology, an 802.11a and an 802.11b module; a dual module is also expected to be available by 2Q03. As for the revenue from wireless communications, Intel expects the ratio to be about 80/20, with 80 percent PC centric and 20 percent communications and networking centric. The expectation is that the wireless and networking capability will grow substantially faster than the microprocessor segment. Most wireless technology comes from Taiwan; however, it is re-branded by the multinational notebook companies, with Askey making modules for Toshiba and Askey, and Gemtek making them for HP and Dell.

7.9.1   Company Overview and Strategy

As a global company, Intel has already invested millions in wireless technologies like Wi-Fi, but the company is also planning ahead to the future of wireless networking much in the same way that it helped develop the Ethernet standard in the 1980s and 1990s. Intel reaffirmed its 802.11

strategy, saying it is looking forward to 2004. That is when the company predicts its "intelligent" wireless roaming will be an established technology, the number of Wi-Fi hotspots should triple, and 90 percent of laptops will have 802.11 wireless LAN capabilities, hopefully with its Centrino chip.

There are challenges, however. Although Intel has already been touting the performance characteristics of the chip, it will have to overcome a marketplace obstacle that it helped create: buyers who judge a chip by its clock speed. Clock speed is measured in megahertz. At 1.6 GHz, Banias will be far slower in terms of raw speed than Pentium 4 notebook chips, which have already hit 2.2 GHz in 2003. Mobile Celerons for the budget crowd already run at 1.2 GHz. Intel's claim is that Centrino will have higher performance and consume less power than today's microprocessors. Notebooks using the chip will consume roughly 25 percent less energy, thereby increasing battery life. Whole subsections of the chip will shut down when not in use to conserve battery power. The general argument for Banias is more work per clock cycle. The Pentium 4 family relies on high clock speed to get performance, and clock speed uses battery power. The public, however, has embraced chip speed as its yardstick, and Intel benefits from this. A successful push on wireless will help Intel sell more processors and increase its wireless LAN IC market share.

7.9.2   Product Overview

The first chips, optimized for laptop systems, are due to appear in the first half of 2003. Systems built on Centrino technology are expected to follow soon after, and should offer firms high-performance laptops with longer battery life than current models.

According to Intel sources, Centrino will feature a large amount of on-chip power-aware cache for high performance at low power consumption, and wireless LAN capability will come as standard. The Centrino, which will eventually replace today's Mobile Pentium 4 chips, will be introduced at speeds of 1.4 GHz to 1.6 GHz, but is expected to outperform current Pentium 4 parts running at 2 GHz. It will have a large 1 MB on-chip cache memory, parts of which can be turned on and off under power management control to conserve power. Wireless LAN capability will be implemented in the Calexico mobile chipset in early Centrino systems, with a dual 802.11a/802.11b solution due in mid-2003. Eventually, the wireless LAN functions may be integrated into the processor itself. In terms of adoption, Intel anticipates the big original design manufacturers (ODMs) like Acer and Asus to be among the first to step up.

Centrino has been designed from scratch to be a low-power, PC-compatible chip and the design incorporates concepts from both the Pentium III and Pentium 4. Part of the power reduction technique stems from the optimized speculation and branch prediction logic, as well as clock gating. Laptops using Centrino technology are eventually expected to boast an eight-hour battery life.

7.10   Philips Semiconductors

Headquartered in The Netherlands, Philips Semiconductor is a leading provider of semiconductor-based solutions for connected consumer and communications applications. Philips Semiconductors' product portfolio includes ASICs, logic, microcontrollers, standard analog, discrete, and power management. Philips Semiconductors has 18 manufacturing and assembly sites, 30 design centers, and four system labs. Philips derives much of its initial strength in the wireless LAN chipset market from its partnership with Agere. The company reports having sold cumulatively more than 6 million 802.11 radio chipsets. Philips offers dual-conversion and direct-conversion transceivers for 2.4-GHz wireless LANs. Another strength of the company is its general analog design and manufacturing expertise. Philips fabricates its latest wireless LAN radio (SA2400A) with its optimized BiCMOS process technology. Wireless connectivity is a strategic focus of the company. Specifically, Philips states that it is working on a complete chipset that includes software and drivers for 802.11b products, indicating that it is expanding beyond radios. Its long-term prospects are better than many other suppliers because Philips is a large, established semiconductor company with close ties to the consumer electronics industry.

7.10.1 Company Overview and Strategy

As a major semiconductor and multimedia consumer electronics company, Philips inherent strength lies in its RF expertise, process, and manufacturing. Its first fully integrated zero-IF single-chip radio, the SA2400, has garnered several significant design wins worldwide. The SA2400 integrates all the functionality needed for radio operation. These include low-noise automatic gain control, fractional-N synthesizers, and transmit and receive filters. In order to provide a more complete solution for next-generation designs, Philips acquired Systemonic in December 2002 in an effort to expand its product portfolio, targeting the highly lucrative multiband market. Systemonic provides technology for a complete dual-band solution, offering a combination of performance throughput, power efficiency, and range that is expected to lead the market. The technology offers better range of up to 70 meters, higher throughput of up to 72 Mbit/s, and lower power usage than most other products currently available. Systemonic developed a compelling configurable architecture optimized for multiprotocol 802.11 applications with the flexibility to extend to future enhancements of the standard. The architecture also incorporates sophisticated power management features that dynamically power-down functions in the chip that are not being used, dramatically reducing power consumption and extending battery life for portable products. Systemonic's products, intellectual property, and systems expertise will add high-speed wireless capabilities to Philips Nexperia SoC solutions.

7.10.2 Product Overview

Philips (Systemonic) develops complete systems solutions that support multiple wireless regional standards, including 802.11a/b/g, HiperLAN/2, and MMAC. The company's chipsets are based on its OnDSP™ platform, which offers a reconfigurable architecture, at the DSP level, to optimize performance to the specific application concerned, resulting in one system for all standards. Philips (Systemonic) currently offers products under the name Tondelayo 2, including a dual-band wireless LAN solution and dual-band radio chipset. Philips (Systemonic) also offers reference boards.

7.11   Marvell Communications

Marvell is a leading provider of semiconductors for broadband communications in the data networking and storage arenas. The company's product portfolio includes switching, transceiver, communications controller, wireless, and storage solutions.

7.11.1 Company Overview and Strategy

Marvell's 2.4-GHz 802.11b wireless LAN solutions are marketed as the Libertas™ product family. The family is divided into three applications: dual-chip PHY layer solution (transceiver and BB), client solution (transceiver and BB+MAC), and access point/gateway solution (transceiver and BB+MAC). All of the products are manufactured using all-CMOS process technology. On February 24, 2003, Marvell announced its "g" Libertas solutions. In August 2002, Marvell introduced its Libertas™ wireless LAN chipsets, including client and access point solutions. The company is currently sampling both chipsets. In January 2002, Marvell announced the sampling of its wireless LAN transceiver (88W8000) and baseband processor (88W8200). The chips are expected to be included in complete wireless LAN solutions.

7.11.2 Product Overview

Libertas 802.11b Client Chipset Solution
For client applications, the Marvell Libertas chipset solution comprises the 88W8000 RF transceiver and the 88W8300 baseband processor/MAC, and includes the following features:

1) On-chip PA with near +20 dBm output power at the antenna port, LNA, variable gain amplifier (VGA), VCO, programmable frequency synthesizer, and RF up/down converters (mixers)

2) On-chip MAC supports 802.11b standard data rates (1, 2, 5.5, and 11 Mbit/s) and Marvell proprietary 22 Mbit/s High Data Rate mode

3) Hardware security engine for WEP and AES encryption and decryption

4) Embedded ARM core with on-chip SRAM memory for low-power operation

7.12 Agere Systems

Agere Systems is the former Microelectronics Group of Lucent Technologies. Agere Systems is a leading provider of IC solutions for high-speed communications and data networks. In October 2001, Agere reorganized its business into two groups: Infrastructure, and Systems and Client Systems. Prior to selling its Orinoco line of business to Proxim, Agere sold end-user products under the Orinoco brand. It also supplies other companies—such as Apple, Avaya (another Lucent spin-off), Dell, IBM, Melco, and Toshiba—with products for sale under their own brand. Agere's first set of wireless LAN products has been based on a multivendor chipset: Philips supplied the RF components, and Agere supplied the MAC and baseband components. To capture additional value, Agere's key advantage as a wireless LAN technology supplier is that it is an incumbent in customer relationships with major system suppliers. Although the organization has a track record of supplying modems, USB controllers, and other components to the PC market, its core business is supplying the communications infrastructure market. Having established its own customer relationships and a sizeable revenue base, the wireless LAN business can probably stand on its own.

7.12.1 Company Overview and Strategy

Not willing to be outdone by its competitors in the lucrative dual-band market, Agere announced its partnership with Infineon Technologies in March 2003. The two companies will enter into a strategic and tactical alliance to develop 802.11a/g, compliant devices. The wireless LAN chipsets, software, and reference designs are expected to start sampling by the second quarter of 2004. The collaboration includes intellectual property licensing and a mutual supply agreement. The companies will independently market the wireless LAN products. Agere will provide its wireless LAN system architecture, multimode medium access controller (MAC), and driver software to the joint venture, while Infineon's main focus will be the development of the dual-band radio technology and wide-band power amplifier chips. The two companies will codevelop the physical (PHY) layer component for the reference designs, combining IP from each firm. Under the deal, the two companies will cross-license relevant patents and technologies to each other.

Another significant event for Agere in 2002 was the sale of its Orinoco product line to Proxim. The sale was made for $65 million in cash. Agere opted to concentrate solely on 802.11 Wi-Fi components. The sale, which includes Agere's Orinoco product line, brings Proxim aboard as an Agere customer. At first glance, this deal has all the makings of a symbiotic relationship: for its money, Proxim is getting a product line that complements the one it acquired earlier via a merger with Western Multiplex. The Sunnyvale, California-based company started by providing high-speed connectivity within buildings; the Agere acquisition now enables it to transition into the last mile. Agere's decision to sell off its systems business was predicated on adding a number of OEM customers like Proxim that are building consumer and enterprise equipment to serve the burgeoning space. The systems piece of the business, while growing, represented only 30% of Agere's wireless LAN business. The problem with the system's business was that it created a strategic conflict with Agere's customers. Chips, modules, and cards, which are sold to systems vendors and OEMs, account for the majority of Agere's business. Of the company's $46 million in system sales in the first half of 2002, components represented about $5 million.

7.12.2 Product Overview

Agere will provide cards, modules, and chips to Proxim. Agere's current offerings include both extended and integrated PC cards, a USB client device, PCI adapters, and reference designs that enable 802.11b standards-based wireless networking.

7.13   Broadcom Corp.

Broadcom designs, develops, and supplies silicon solutions for broadband networks. The company's systems and component solutions are implemented in set-top boxes, cable/DSL modems, residential gateways, high-speed transmission and switching for local, metropolitan and wide-area networking; wireless SOHO and cellular networks; VoIP gateway and telephony systems; and broadband network processors. As a formidable player in home networking, Broadcom has the wherewithal to take on industry leaders Intersil and Atheros.

7.13.1 Company Overview and Strategy

As a company that is laser focused on getting design wins, Broadcom's achievement to date has been impressive. On March 12, 2003, Broadcom announced that its 802.11g and 802.11a/g products are now being offered as part of Dell's Latitude D-Family notebooks. On March 10, 2003, Broadcom announced it shipped about 1 million "g" chipsets on a cumulative basis. In July 2002, Broadcom announced its new three-chip 802.11a/b wireless LAN chipset. The company is currently sampling the solution to early access partners. Because the software implementation on both its 802.11b and dual-band solution is the same, Broadcom expects migration from its existing installed base of 802.11b customers to this new chipset. In January 2002, Broadcom announced its entry into wireless LAN market with the introduction of its 802.11b chipset. Broadcom also introduced a line of 802.11b products in January 2002. The line includes a direct-conversion radio implemented in CMOS and several baseband/MAC chips. One baseband/MAC chip is a basic model; the others add Fast Ethernet interfaces and an analog modem. Details have not been disclosed. Based on their configuration, the products appear to be targeted at the NIC market. The inclusion of conventional LAN and modem capabilities will enable the construction of multifunction PC cards or mini peripheral component interconnect (miniPCI) cards. Including modem capability is especially sensible. Modems, because they connect to public infrastructure, are regulated and approved by each country. Therefore, they are commonly included in notebook PCs as add-in cards.

Similar regulations apply to wireless LAN cards. It therefore makes sense to integrate the two functions on the same card. Broadcom has the competitive advantage of an established presence in the LAN and broadband IC markets. It has relationships with major suppliers of NICs (except perhaps Intel) and other LAN equipment. Broadcom also has established customer relationships with several PC companies, including Compaq and Dell. Further, Broadcom is a large supplier of cable modem and set-top box (STB) chips. Potentially, the company can design wireless LAN functions into these products; at a minimum it can take advantage of the existing marketing channel. Broadcom also markets Bluetooth chipsets and is therefore well-positioned to develop combination wireless LAN–wireless PAN chipsets.

7.13.2 Product Overview

Broadcom's wireless product portfolio consists of wireless LAN chipsets for Bluetooth, 802.11a/b, and direct broadcast satellite applications. The BCM94301 reference design family supports 802.11b and is based on the company's BB+MAC (BCM430x) and radio chip (BCM2051) for incorporation in NICs. In Q103, Broadcom announced sampling for a three-chip, dual-band 802.11a/b chipset (BCM94309MP) based entirely on a CMOS process. The three chips are BB+MAC (BCM4309), 2.4-GHz radio (BCM2050), and 5-GHz radio (BCM2060).

7.14   Future of Wireless LAN IC Industry

The future looks bright for the wireless LAN IC industry. Ultimately, the key decision factor that determines the success of a wireless LAN IC company is the number of design wins, which translates to having the largest market share. To that end, BOM cost, radio performance, and features are all important differentiators for wireless LAN IC suppliers. Suppliers that can reduce the time it takes to bring to market products with new features such as 802.11g compatibility and WPA while reducing BOM cost and maximizing radio performance will have an advantage in securing design wins.

7.15 Summary

The phenomenon that will constantly plague wireless chipset companies is price erosion. The average sales price (ASP) for 802.11b is expected to fall below $7.50 by the end of 2003. Many factors contribute to this erosion. Taiwan Inc. will be a major force in cost reducing their offerings. With their ultra-low cost baseband/MAC and RF from Philips and other RF companies, they will create drastic competitive pressures for the incumbents. Another factor that will drive the price curve down is the emergence of tri-mode and 802.11g-only solutions. These solutions will become mainstream for the computing market in 2003 and beyond. Finally, continuous improvements in the IC manufacturing process, embedded designs, and Centrino effect will lower the overall pricing for all flavors of 802.11x.

Chapter 8

Emerging Trends and Case Studies for Wi-Fi

The wireless LAN market is one of the fastest growing markets today. Over the last three years, a very rapid wireless evolution has occurred in the home and the public space. In 1999 and 2000, HomeRF was considered a strong contender for the technology of choice for the home. In 2001, HomeRF dropped out and Wi-Fi became the natural choice for the home. In 2001, the main issue was whether 802.11a or 802.11g would be the technology of choice for the home and the enterprise. In 2002, the wireless home landscape broadened, with a handful of wireless technologies showing significant potential for specific applications in the home. These include UWB, Bluetooth, and IEEE 802.11. The wireless LAN chipset market is extremely competitive. The barriers to entry have been significantly reduced since the standards received approval. The IEEE 802.11 chipset market, especially 802.11b, is the most competitive and the most successful. Although Intersil still commands the lion's share of the 802.11b market, the emerging dual-band IEEE 802.11a market will be more competitive, even for the incumbent IEEE 802.11a leader, Atheros. Uncertainty over which standard to support has caused many vendors to develop dual-mode 802.11a/g solutions. Some of the 802.11 chipmakers are also integrating Bluetooth in their designs.

8.1   Major Trends for 2003

Overall, 2002 was a good year for Wi-Fi. Looking ahead, here are the major trends for 2003 and beyond:

- Wireless interworking with cellular

- Price erosion to continue
- Increased competition
- Impact of pre-802.11g on the marketplace
- High throughput wireless LANs
- Low power 802.11

8.2   Wireless Interworking with Cellular Networks

Considerable interests have spurred the development for products that played a part in the convergence between wireless LANs and cellular technologies. For example, Nextel and Motorola announced at CTIA 2003 that they are working on a mobile phone with integrated Wi-Fi capability, which will allow users to make calls over a Wi-Fi home or office network and use the same phone to call over the network while on the road. Although it cannot truly roam from one type of network to the other, this phone could allow customers to replace a cordless phone or a wired PBX (private branch exchange) with a wireless LAN. A call initiated on the network will stay on that network until the user gets to the office, but a call started on the Wi-Fi network will be dropped once the user moves out of range. Conceptually, Nextel and other providers can also work with providers of cable data services to send phone calls from the Wi-Fi network over the cable infrastructure instead of the traditional telecommunications network. Therein lies another business proposition for cable access to the home.

True handoffs of voice calls between Wi-Fi and cellular networks pose big technological problems because Wi-Fi technology is designed to be inexpensive and relatively simple. Cellular technologies have in-built special standards to allow handoffs as users move around, and those specifications were not built into the IEEE 802.11 standard on which Wi-Fi products are based. This is about to change, however. In traditional cellular networks, handoff of a terminal from one base station to another is a critical function. Since such a handoff is handled primarily at network Layers 3 and 4, it is not directly supported by IEEE 802 standards, which specify only Layers 1 and 2. As handoff is becoming increasingly important for IEEE 802 wireless standards, the IEEE 802 has started exploring the general issue of handoff and the means by which it is achieved in cellular networks and in IP Mobility specifications. Considerable discussions have also centered on the means by which IEEE 802 standards might interface with higher-layer mechanisms and thereby support handoff specified at higher layers. Since the focus is on the direct support of higher-layer functionality, the general approach is that a single IEEE 802 handoff interface may be implemented for all IEEE 802 devices. This will allow handoffs among mixed IEEE 802 networks and non-IEEE 802 systems. An important activity resulting from these pockets of interest has culminated in a series of efforts to harmonize the activities of IEEE 802.11, ETSI, and MMAC. The primary motivation for this global coordination stems from the following:

- The need to present a generic interworking interface to cellular networks and external IP networks from the wireless LAN standards

- This interface carries a multitude of functionalities covering broad areas like accounting, security, handoffs, billing, resource measurement, and Quality of Service.

On the standards front, the IEEE 802.11 Working Group recently formed a new activity called the Wireless Interworking Group (WIG) with ETSI BRAN and MMAC. The main objective of WIG is to establish a joint-effort between IEEE 802.11 and ETSI BRAN/MMAC HSWA for the interworking of wireless LANs to 3G cellular systems. WIG had its first meeting with participation from ETSI BRAN and MMAC at the September 2002 meeting in Monterey, California. At the Monterey meeting, the group reached an agreement on the scope, administration and organizational issues, work items, and production of official statements. In addition, WIG developed liaison statements to 3GPP SA, 3GPP2, GSM Association, Wi-Fi Alliance, PassOne, IETF, The Open Group, and HiperLAN/2 Global Forum. The purpose of these liaison statements is to inform of the existence WIG and also establish procedures for exchange of documents. The official scope of WIG is as follows:

- To be an integral part in the production of a widely applicable interworking standard for Wireless Wide Area Network (WWAN) and other public networks. The standard is to be applicable for IEEE 802.11, MMAC HiSWAN, and ETSI HiperLAN/2.

- To be the point of resolution for ETSI, IEEE, and MMAC on issues related to interworking with WWAN and other public networks.

- To be the single point of contact for the above-mentioned wireless LAN standards on questions related to interworking with WWAN and other public networks.

The ETSI BRAN group started interworking activities back in mid-2002 and to some extent is the victim of its own success. It has achieved what it originally set out to do. It has defined a set of requirements for wireless LAN interworking and produced a suitable engineering solution to enable development of a prototype system. Over the past year or so, however, the enthusiasm within ETSI BRAN has waned. Since the start of WIG and the production of the WIG baseline document [1], the majority of the activity has shifted to the IEEE 802.11 Wireless LANs Next Generation Committee. By now it is clear that HiperLAN/2 will not be a major wireless LAN technology even in Europe. It appears that most companies seem to have left ETSI BRAN to follow either 3GPP SA or IEEE 802.11 for future work on wireless LAN interworking with 3GPP and external IP networks. This momentum has carried through within IEEE 802.11 and 3GPP SA. The IEEE 802.11 now has wireless LAN Interworking as an important activity under the WNG SC umbrella. 3GPP SA have also decided that there are certain aspects of interworking that can only be addressed by the IEEE 802.11. In March 2003, the IEEE 802 Working Group also approved the formation of a study group to evaluate the possibility of developing a standard specifying a common handoff framework applicable to 802 standards. Specifically, the efforts will focus on the following:

- Mechanisms for supporting roaming decisions either within a mobile device or within network devices (e.g., APs or base stations)

- The type and structure of relevant information that may be passed between entities within a network

- Maintenance of compatibility with existing related mechanisms (for example 802.11k[1], 802.11f, and 802.16 handoff)

- Issues surrounding interworking with upper-layer protocols (e.g., IPv6, IPv4)

- Mechanisms for establishment and maintenance of security within a roaming context

- Layer 2 mechanisms to support seamless roaming

- Mechanisms for supporting scaleable roaming and mobility architectures

- Summary of L2/L3 handoff schemes in 3GPP networks and/or other IP wireless networks

- Requirements for supporting roaming of bursty and real-time applications

- Outline descriptions of functional components for active and dormant mode handoffs within and across 802 wireless technologies

8.3 Price Erosion to Continue

The wireless LAN market is one that is becoming increasingly commoditized. To aggressively capture market share, Taiwan companies will be prepared to offer single-

---

[1] The 802.11k Task Group defines Radio Resource Measurement enhancements to provide interfaces to higher layers for radio and network measurements.

chip solutions at price discounts of more than 30% to global players like Intersil. Players with in-house capability will be the ones who can compete on price.

8.3.1   Case Study — Cost Reduction Strategy

The following section provides a case study on how a company can reduce cost in producing a chipset for the dual-band market. The challenge here is determining the market entry price. As this is a moving target, it becomes imperative for the company to continually monitor price erosion and prepare for these dynamics in the marketplace.

Company A is in the business of providing reference designs to OEMs and ODMs. They have a dual-band chipset that they hope will capture 10-15% of the total addressable market (TAM). Their share of available market (SAM) is about 5% for the first year, increasing to 10% for the second year, and targeting 20% from year three onward. However, the price erosion for dual-band chipsets is expected to be even more severe the year the product is launched. This is due to the number of new entrants into the market. Even the incumbents are prepared to accept a smaller margin in order to gain some market share. With price pressures being applied from customers to suppliers, it became imperative for Company A to decide their market entry price. Based on discussions with customers and partners, they established a price erosion curve relative to their projected BOM cost from 2003 to 2006 (Figure 8.1).

From the analysis, it became obvious Company A will not be competitive. Management decision is clear at this juncture. They had two options. They can forget about entering the market because the Return On Investment (ROI) will not justify spending $15 million per year on R&D and engineering to develop a solution that was not price competitive. For option 2, they can look into cost reduction strategies. They told engineering to look at tradeoffs they can consider to further reduce the cost of building the chipset. Part of this task involves identifying the shortcomings of the present solution, doing a gap analysis and presenting both long and short-term options.

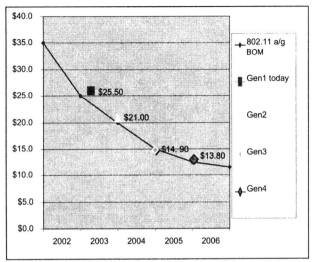

Figure 8.1: BOM Price Erosion based on Current Configuration.

After careful consideration and further analysis, engineering came back with various options. Marketing engaged in various discussions with customers and identified price points that the market can bear. Engineering analyzed each component of the BOM and discovered areas where they can further reduce the cost of components. In some cases, it appeared more cost competitive if there was a possibility of porting to a better IC process, whereby the die size can be further reduced. This will result in a lower BOM. From the analysis performed, they were able to cut the cost of components drastically, resulting in a BOM of $23.50 as an entry price in 2003. This was an important milestone as it allowed Company A to introduce their products in 2003 and be within the top-three position from a price-competition stand-point. If they had to reengineer the chipset, in particular the radio, they would have lost valuable market time.

Essentially, they were able to reduce cost by focusing on the rest of BOM, which includes the power amplifier (PA), band-pass filters (BPF), and replacement of expensive filters and analog-to-digital converters (ADCs) with cheaper alternatives such as ceramic filters, and sourcing for ADCs internally. In order to sustain the price erosion, Company A had to present a series of targets they felt were achievable based on current and future technology. For example, one of the key decision criteria involved the optimization of the baseband. The baseband as it stood was priced at about $6.00. Through further analysis, it was determined that if a port were made with a newer process, it would be possible to shrink the die size, thereby resulting in a lower BOM cost. Even the radio section could be further improved. The generation 1 radio is a superheterodyne architecture and includes an external oscillator. The number of external components is high. Based on this information, the team came back with a revised chart for target BOM costs relative to the price erosion expected in Figure 8.2. This figure demonstrates the generation 1 subsystems with a total BOM of $23.00. The key challenge for the engineering team was keeping the baseband and RF unmodified. Most of the optimization is in the external BOM count.

Another area that can be further optimized is the PA. Most leading vendors have tried to integrate the PA into the RF front-end, thereby reducing the overall BOM cost. However, the performance resulting from this integration is inadequate. The integration remains a challenge for most vendors. This is because the PA in 802.11a has very high peak-to-average ratios, which demand very linear designs. Although these designs are doable, they normally involve expensive technologies and require a lot of power to operate. The challenge here is to implement a low-cost design using CMOS or other nonexotic materials with adequate fidelity while maintaining power efficiency. Many companies resort to cost reduction of their chipset through the use of extensive integration and the implementation of a zero-IF (ZIF) radio transceiver on CMOS. Integration leads the way to lower BOM in many ways.

In general, a high degree of integration can lead to a number of positive results. They are outlined below.

- Shrinking form factors due to reduced component counts

- Quicker time-to-market cycles, thus beating the competition in terms of consumer purchasing

- Less requirement for major in-house RF expertise due to packaged RF modules

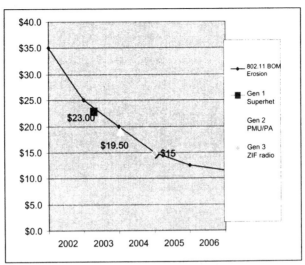

Figure 8.2: Revised Target BOM Relative to Price Erosion.

- Fewer supply sources, which translates into easier supply chain management

- Avoidance of standards certification because of pre-certified solutions

- Fewer integrated parts, resulting in a reduced BOM

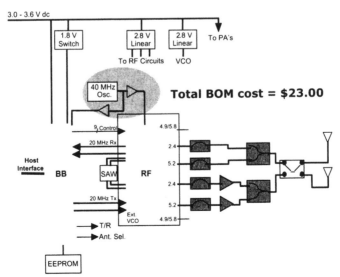

Figure 8.3: BOM for First-Generation Chipset.

Using module technology to provide integration, Company A was able to develop an RF front-end module combining the PA and switch capabilities. This resulted in a lower component count. Because the module technology was fairly well understood, they were able to further reduce the BOM as well. Another area of integration was in the power management unit (PMU). With these additions, the overall BOM was further lowered to $19.50, as indicated in, Figure 8.4.

All these techniques were centered on the rest of BOM. They had not modified the baseband and RF portion of the entire system. Generally, further reduction may be obtained by adopting a ZIF architecture for the radio. The approach taken here was to evaluate other alternative receiver architectures with the goal of determining the architecture best suited to scalability, cost, and performance. The super-heterodyne architecture is still widely used in the industry and is very well understood and robust. However, due to the cost-sensitive nature of wireless LAN IC chips, it is being replaced by newer architectures such as ZIF and sliding IF. Figure 8.5 shows how the final target BOM of $15 is achieved. Basically, the change is in the radio architecture from a superheterodyne to a direct-conversion ZIF, resulting in fewer components, smaller size, and lower power consumption.

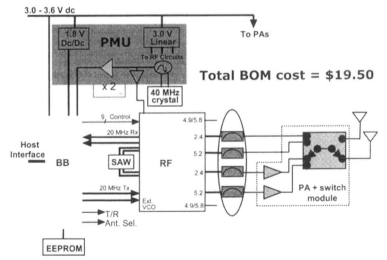

Figure 8.4: Improved BOM in Second-Generation Chipset.

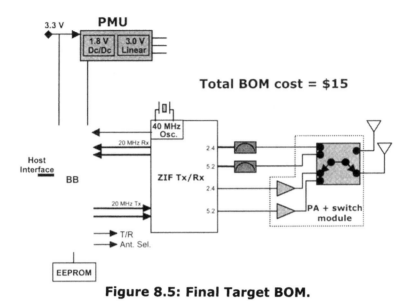

Figure 8.5: Final Target BOM.

There is no free lunch, even for the ZIF architecture. There are well-known problems with the ZIF approach, in particular the dc offset and I/Q imbalances. Because the radio is designed to be a dual-band radio, the problem is exacerbated due to the more stringent I/Q imbalance and dc offset cancellation time requirements for the OFDM receiver. These and other challenges such as AGC settling time and frequency-dependent I/Q phase mismatches need to be overcome.

8.4   Increased Competition

One of the most difficult decisions corporations have to deal with is the execution of the strategies and staying on top of the competition. Very few companies in the wireless business have the luxury of being the only company with a product that is in high demand. It is, however, relatively easy to determine if your company's gross margin is in line with your competitors.

If you take the top three competitors in your business and compare their financial numbers, very often you can tell if you are operating within the ballpark or away from the average. For example, take the case of the low-power Wi-Fi market. The top three competitors as of Q1 2003 are TI, Marvell, and Intersil. By doing a scan of their yearly report, such as their 10-K, or quarterly report, such as the 10-Q, you can get a pretty good sense of the level of competitiveness based on their estimated revenue. Also, you can compare financial targets of these companies by looking at their gross margin (%), R&D budget as a percentage of sales, marketing as a percentage of sales, and their income from operations (IFO). The IFO is also known as the operating income. Also, public companies such as Intersil, TI, and Marvell have many product lines of business. Sometimes, you can get the breakdown by product groups that can be useful in terms of gauging the overall importance of the segment of business you are analyzing. Table 8.1 illustrates the numbers for the three companies.

The gross margin is an important figure for basic analysis. It gives a rough guide to the percentage typically used by the company in pricing its products. Knowing the cost to produce the product and other relevant details can quickly give some guidance on the BOM of your competitor's products. Another interesting figure is the R&D percentage of sales. In this market, where engineering innovation can be a key differentiator, knowing how much your competitor allocates to R&D spending can provide some insights into their engineering capabilities.

Finally, a look at the IFO can clue you in on the overall health of the company and how it plans to compete in the future. Clearly, when combined with market data for Wi-Fi equipment or chipsets shipped, one can determine quickly if a company is losing market share and to whom the majority of sales is going.

Table 8.1: Financial Ratio Analysis of Peer Group.

	Intersil	Texas Instr.	Marvell
Revenue (2002) M$	650	8383	505
Revenue (2003)	719	9300	726
Gross Profit/Sales	51.88%	36.60%	53.88%
Operating Income/Sales (IFO)	0.68%	3.44%	-14.05%
R&D (% of sales)	20.17%	19.30%	28.84%
Marketing, Sales and Gen Adm (% of sales)	17.20%	13.90%	12.43%

8.5 Impact of Pre-802.11g Products in the Marketplace

The hype surrounding the 802.11g specification has been building just before the Comdex 2003 show, with both chip developers and OEMs touting their 802.11g solutions. As 802.11g gains steam, questions are also surfacing about potential interoperability problems between existing 802.11g systems on the market and those that were just released.

The fact that 802.11g is still at the preapproved stage in the beginning of 2003 did not stop several wireless LAN IC companies from releasing "802.11g-ready" chipsets, most notably Broadcom, Intersil, and Marvell. These suppliers have buyers ready to roll out 802.11g solutions to the home and enterprise. Linksys, D-Link, and Buffalo are among the first few. Even Apple Computer has stepped into the ring, announcing its new line of Airport Extreme, which it states is based on the "cutting-edge" IEEE 802.11g wireless draft specification.

8.6 High-Throughput Wireless LANs

The very rapid growth of wireless LANs beyond traditional, low-bandwidth, vertical applications (primarily involving data collection and similar activities) and into mission-critical general-office applications began to build with the approval of the initial IEEE 802.11 standard in June 1997. However it was not until the advent of the 11-Mbit/s 802.11b standard in September 1999 that the horizontal wireless LAN market achieved some semblance of legitimacy, followed by rapid acceleration. Long anticipated by many analysts, the current fever surrounding wireless LANs is a bright spot in an otherwise lackluster networking market. Worldwide sales in 2003 based on the level of activity in the major wireless LAN submarkets will be on the order of US$1.9 billion. And yet, even with this level of success and growth, wireless LANs are not immune to the rip currents that pervade essentially every high-technology market.

The dominant factor is the demand for improvements in price/performance. It comes as no surprise, then, that wireless LAN vendors have been sparing little effort in building additional tools and now entirely new architectures to address the above needs. It is this drive that led to the IEEE 802.11 Working Group creating a Task Group, namely IEEE 802.11n, to look into the issues related to increasing the throughput of existing wireless LAN systems. The charter for 802.11n is to create an amendment that shall define standardized modifications to both the 802.11 physical layers (PHY) and the 802.11 Medium Access Control Layer (MAC) so that modes of operation can be enabled that are capable of much higher throughputs, with a maximum throughput of at least 100 Mbit/s, as measured at the MAC data service access point (SAP). The SAP is the point at which the services of an OSI layer are made available to the next-higher layer. In this case, it is the interaction between the MAC and Logical Link Layer (LLC). The purpose for this activity is to improve the wireless LAN user experience by providing significantly higher throughput for current applications and to enable new applications and market segments.

8.6.1 Market and Applications

To really understand the market demand for high-throughput wireless LAN, it is useful to learn from history how technologies evolve. In the early 1970's, Ethernet emerged as a networking protocol designed to link computers throughout a building and campus. Since those humble beginnings, Ethernet has evolved into 10Base-T (IEEE 802.3) in 1983, Fast Ethernet (IEEE 802.3u) in 1995, 1-Gigabit Ethernet (802.3z) in 1998, and 10-Gigabit Ethernet (802.3ae) in 2002. Thus, developments in the evolution of Ethernet have come more rapidly and have entailed a great expansion in capabilities.

Today, Ethernet not only dominates the LAN market, it is also taking hold in the MAN and is extending into the WAN, as both its distance and capacity increase. Not surprisingly, Ethernet at the 1-Gigabit and 10-Gigabit levels has attracted increasing attention from carriers operating in the MAN and WAN. However, the full promise of Ethernet in the MAN and WAN has yet to be realized, as 1- and 10-Gigabit Ethernet technologies have not yet been fully exploited. In the access market, Ethernet in the form of Ethernet passive optical networks (PONs) has also been minimally deployed. The key in technology evolution has always been capacity. The need for higher bandwidth is still the key driver toward technology adoption. In the case of wireless LAN, the same holds true. If we plot the chart of evolution in wired Ethernet versus wireless Ethernet (see Figure 8.6), it is interesting to note the similarities in the rate of adoption of higher speeds in both wired Ethernet and wireless Ethernet.

Today the most common work performed across wireless LANs involves office applications, such as e-mail, spreadsheet building, Web browsing, and word processing. Both 802.11b and 802.11a can handle such traffic. But as streaming video and dynamic digital content become more common, the throughput of 802.11b products will not suffice. Clearly, the present state of Wi-Fi will not satisfy all the requirements for AV streaming within the home today. However, with the ratification of 802.11e and 802.11i, we are a few steps closer in meeting that reality. This bodes well for IEEE 802.11n as it promises to provide, as a minimum, a theoretical throughput of at least 100 Mbit/s.

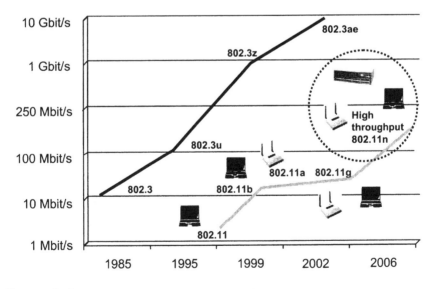

Figure 8 6:   **Data Rate Progression in Ethernet and Wireless LANs.**

Today, the answer to AV streaming is 802.11a. Ironically, this standard was approved shortly before 802.11b, but the technology presented more engineering challenges, and products did not enter the market until late 2002. 802.11a products operate at a higher frequency and have a higher throughput, but at a shorter range. An added advantage of 802.11a for a relatively large organization is the number of channels available. The 5-GHz spectrum has more bandwidth allocated for 802.11 wireless technology than the 2-GHz spectrum. As a result, the 5-GHz 802.11a standard has eight nonoverlapping channels available—five more than 802.11b technology. With more channels, more access points can be added to an area without interference. Still, 802.11a has some issues to be worked out, particularly in the area of compatibility. For starters, 802.11a products are not backward-compatible with 802.11b products, which clearly dominate the market. Table 8.2 outlines the bandwidth requirements for supporting various applications in the home and enterprise. From Table 8.2, we can deduce that applications requiring multiparty communications will likely suffer in performance due to insufficient capacity.

As video conferencing becomes more ubiquitous in the workplace, it is apparent that to support a wireless infrastructure consisting of wireless clients and wireless servers, a high-speed wireless backbone will be a standard feature in the office networks of the future. This high-speed wireless network will likely be an IEEE 802.11n-compliant network, interconnecting clusters of IEEE 802.11a/g or IEEE 802.11b devices. For that same reason, it is possible to expect ultrawideband (UWB) devices to provide the intra-cluster communications and IEEE 802.11n devices to provide the intercluster communications. Therefore, the emergence of UWB-like devices will highly complement high-throughput wireless LAN in a multicluster office environment.

Even though most market studies suggest that major growth is expected in the home, with consumer electronic devices becoming more and more Wi-Fi capable, the actual rollout of wireless networking may be different. It is believed that the next area of growth for high-throughput wireless LANs, is in the large enterprise. In fact, with the current global economic weakness, it is foreseeable that the major IT spending will still be carried out by the large enterprise.

Table 8.2: Bandwidth Requirements for Typical Applications in the Business and Home.

Devices	Applications	Typical Bandwidth Requirements
MP3 player	Audio file transfer	60 – 320 Kbit/s
MPEG4	Video	750 Kbit/s
Notebooks/Desktops	E-mail, Web browsing	1 Mbit/s
Notebooks/Desktops	Video conferencing	1.5 – 3 Mbit/s
DVD	Video burn-in	9.8 Mbit/s
HDTV	Video	19.39 Mbit/s

The consumer electronics industry will also be important but the process will be gradual. From a short-term perspective, it is anticipated that corporate IT spending will continue to carry the demand for higher-throughput wireless LAN technologies.

Apart from the home, wireless LANs have also found an unusual application in storage area networks. The first wireless IP storage area network was developed to produce cost savings. Nishan Systems, XIOTech Corporation, and Proxim combined their respective expertise to build the WSAN at Steinbech Credit Union in Manitoba, Canada in early 2003. The WSAN speed, with a transfer rate of 7 Megabytes per second, allowed Steinbech to back up all 6 terabytes of its data, not just the critical stuff. To be successful in AV streaming, wireless LANs are expected to offer unprecedented link reliability and throughput between broadband network devices and consumer electronics products, as well as home computing platforms.

8.6.2   Role of IEEE 802.11 in the Home

Traditionally, home networking technologies have been confined mostly to Ethernet platforms. In fact, some of the earlier home development projects touted the availability of Fast Ethernet as an attraction to potential buyers. Of course, what is provided here is basically Cat 5 cabling and, depending on how network-capable the home is, some homeowners only need to plug in their desktops and printer to create the home network. These days, the trend is moving quickly to other options. There are a variety of technologies that can offer compelling benefits. These are the IEEE 802.11, IEEE 1394, HomePNA, HomePlug, and UWB.

The market for 1394-enabled digital products has been steadily expanding both in the CE and PC domains. Over 15 million 1394-enabled PCs and PC peripherals and over 7 million 1394-enabled CE devices were sold in 2000 alone. There are currently hundreds of 1394 products on the market, including camcorders, DTVs, STBs, MD players, PlayStation 2, digital video recorders, video editing systems, music synthesizers, speakers, desktop PCs, notebook PCs, printers, scanners, HDDs, digital cameras, PC cameras, Zip drives, CD-ROM drives, DVD+RW drives, MO drives, DAT drives, RAID systems, etc. The market size for 1394 chips was $180 million in 2000 and is expected to grow to $590 million in 2005. As the market for 1394-enabled products expands, clusters of 1394-enabled devices are forming around the home, and there are needs for enabling room-to-room communications among these clusters to build an in-home digital network. Wireless is an attractive solution to this problem because of the flexibility and ease of installation it provides.

For wireless 1394 products to gain popularity, certain conditions need to be met. First, there should be a stable wireless technology that supports isochronous (real-time) streams. Second, there should be bridge-aware devices (and protocols) on the market that allow seamless communications among devices in a hybrid wired/wireless environment. The situation that will prevail however, is the technology that guarantees compatibility and inter-operability. For the most part, if the office at work uses wireless Ethernet, then more than likely, the choice at home will be the same. This means lower maintenance and less technology headaches to deal with. There has been a flurry of activities in the IEEE 802.11 domain to address the needs of the home. Various scenarios have been articulated with a variety of technologies. The good news is there appears to be a convergence toward wireless LAN as the platform of choice.

While there are many competing technologies such as antenna diversity that can potentially provide higher throughput, the one technology that seems to have most widespread support is Multiple-Input-Multiple-Output

(MIMO). Many research studies have indicated that systems can achieve more than 100 Mbit/s using MIMO-OFDM with a small performance degradation. For example, in [2], the authors concluded that with BPSK and QPSK modulation, they can achieve better performance than with Single-Input-Single-Output (SISO). With 16- and 64-QAM, performance degradations due to MIMO are about 0.5 and 2 dB, respectively. They also concluded that high spectral efficiency can be achieved by using MIMO-OFDM; hence, MIMO-OFDM is a good potential solution for IEEE 802.11n.

Applying the concept of multicluster communication in the home, it is highly probable that the digital home of the future will consist of multiple clusters. These clusters could fall into four broad categories: Entertainment, PC/Productivity, Communications, and Home Control and Management. The idea of cluster communication is to allow consumers to connect fragmented clusters of digital devices connected by a variety of technologies such as 1394 and Bluetooth. Most of those clusters will be confined within a room due to the physical limitations of the personal area networks such as UWB, Bluetooth, or IEEE 1394. For instance, there may be a living room cluster consisting of a digital STB, PlayStation 2, and a digital VCR, and a home office cluster consisting of a PC, a printer, and an external HDD. A wireless bridge device that is added to each cluster will enable communications between these physically separate clusters without needing new wiring. This wireless bridge will likely evolve out of the IEEE 802.11n/e/i standard, incorporating the high-speed capabilities with full quality of service and security support. For instance, the user will be able to watch a TV program on his/her PC, or use the external HDD with a time-shift device such as the TiVo. Also, wireless devices can be added to the network. For instance, a PDA-like portable device will allow the user to move around the home while controlling the digital devices in the home, and also watch a video/TV program on its small screen. Another example would be a wireless set-top box (STB) or video server distributing video streams to wireless video monitors throughout the home.

8.7   Low-Power 802.11

Chapter 6 provided an overview of the wireless LAN market and where it is headed. It covered several key areas in the hardware and wireless LAN IC industries and the trends that follow. The adoption of Wi-Fi has spread beyond computing devices. In particular, it has moved quickly to PDAs, converged phones, and other devices that provide basic mobility to the owner. Having the ability to connect wirelessly through Wi-Fi has spawned a new breed of applications and products. This is the next high-growth area for Wi-Fi today. This section will cover the issues and trends related to the Wi-Fi-cation of devices that are sensitive to power consumption and compactness of solution.

The market drivers are beginning to emerge at wireless trade shows and exhibitions. At the CTIA show in New Orleans (March 2003), Nextel Communications said they and Motorola will offer a mobile phone with Motorola based on Microsoft's Windows-Powered Smartphone platform, and in the second quarter of this year the carrier will begin trials of a phone that can be used on Wi-Fi wireless LANs. Also, Texas Instruments demonstrated a concept design for a Pocket PC Phone Edition device that supports three modes of wireless connectivity: GSM/GPRS cellular, 802.11b Wi-Fi, and Bluetooth. Who needs it? Imagine someone with a Bluetooth headset, looking at the PDA while talking on a GSM call and browsing the Web via 802.11b. Another application, which will lead to further convergence of computing and communication, is remote access. ExpertCity, makers of the remote-access service GoToMyPC, are introducing GoToMyPC PocketView, a version of the service that runs on Pocket PCs. GoToMyPC PocketView is a real-time remote-access solution that enables users of wireless devices to view and control their PCs in real time. Targeted mainly for mobile professionals, this application provides fast, easy, and extremely secure access to all of the data and applications on users' desktops from handheld devices. This includes access to all software and network connections, including all proprietary corporate applications wherever they are, instead of just e-mail, calendar, and a few "Webified" programs, resulting in greater efficiency and productivity. All this points to a greater demand for mobile devices (smartphones, converged phones, PDAs) to embrace Wi-Fi and this presents a challenge to the chipsets and interfaces, where battery consumption and compactness become important.

To start this discussion, we consider what devices need power-saving capability. Contrary to popular belief that the only notebooks PCs and laptops are the main drivers for power-saving devices, there is currently an emerging market for 802.11-enabled SMART cellphones, PDAs, and in the very near future, 802.11 VoIP-based cordless phones. The key challenge here is minimizing the Wi-Fi burden on user expectation of battery life. This is especially crucial for Windows CE types of devices. In addition, the need for smooth cellular handoffs with VoIP in Wi-Fi hotspots again requires proper battery power management in order to sustain the duration of the handset. From the standards perspective, the original IEEE 802.11 standard issued in 1999 provides a power management mechanism that allows mobile stations to "sleep" without missing any packets. In the latest draft of the IEEE 802.11e task group, the draft 802.11e version 4.0 provides two additional power-management mechanisms that allow QoS-capable mobile stations to "sleep" without missing directed packets.

8.7.1 State of Power Management in Wireless LAN Today

Most wireless LAN clients will be portable systems. The fact that portable systems are typically battery-powered led to another area of specialization in the 802.11 MAC. Specifically, features were added to the MAC that can maximize battery life in portable clients via power-management schemes. When used in infrastructure mode, the user of a mobile 802.11b device can decide when to connect to a nearby AP and the conditions under which to maintain connectivity. There may be portions of the day when the user is out of range of a wireless network, or chooses not to be connected, such as when commuting or after business hours. These connectivity preferences can be stored in a device such as a PDA and used to optimize battery life.

To support clients that periodically enter sleep mode, the 802.11 protocol specifies that APs include buffers to store messages. While connected, the device can be placed in a power-saving "sleep" or standby mode between beacons, those instants in time when the 802.11b subsystem looks for information from the AP. Beacons are transmitted at precise intervals, 15 times per second for example, and are used by the wireless LAN to identify all network members, and to alert these stations when data is waiting to be transmitted to them. This ability of Wi-Fi devices to enter a sleep mode enables a reduction in average power consumption. Sleeping clients are required to be awakened periodically to retrieve any messages. The APs are permitted to discard unread messages after a specified time, and the messages go unretrieved.

Figure 8.7 illustrates the interaction between the client and AP. The client does not have to be awake for every beacon. The example in Figure 8.7 shows the station waking up at every second beacon. The 802.11 protocol allows the station to use a parameter called the Listen Interval to save additional power. The Listen Interval is a parameter sent to the AP during network connectivity. The use of longer Listen Intervals allows the device to miss a specific number of beacons without losing any data traffic or disconnecting from the network. By using the Listen Interval mechanism, the AP buffers data while the station is asleep and not listening. The Listen Interval can be set at 10, for example, so that when 10 beacons occur per second, the device will wake up and listen once every second. The device's response time using this Listen Interval would not be perceptible to users. The system would function as if it were continuously connected to the wireless LAN. It is important to note, however, that the AP and network must be able to handle data-buffering requirements for all associated devices.

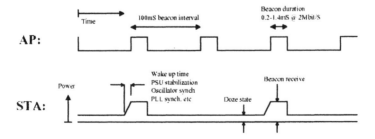

Figure 8.7:   **Interaction between Client and AP in Power Save Mode.**

The power-management scheme is relatively easy to implement in a Basic or Extended Service Set configuration because the AP is always present. Moreover, the AP is never battery powered and never enters sleep mode. In an Independent Basic Service Set (IBSS) configuration, however, no AP exists and all devices may desire to enter sleep mode. The 802.11 protocol does provide for power savings in an IBSS configuration. Essentially, all clients in an IBSS configuration must be awaken each time a beacon is sent. The clients randomly alternate the task of transmitting a beacon. Immediately after the transmission of each beacon, a short time period, called an announcement traffic information message (ATIM) window, commences. During the ATIM window, any station can indicate the need to transfer data to another station during the ensuing data-transmission window. Clients with no incoming or outgoing frames pending can reenter sleep mode during the data-transmission window.

8.7.2 Trade-Offs in Power Consumption

In general, there are two important factors that affect power consumption in a mobile 802.11b device. These are the range supported and the transmission rate between the AP and the stations. The relationship between data rate and power consumption is more complicated. For example, it is quite intuitive to assume that if we restrict operation to lower data rates (which are less susceptible to interference), in combination with lower RF transmit power, we can reduce the 802.11b power consumption. The same usable range or distance for the device could be maintained with reduced transmitting power because lower data rates are more tolerant of interference.

The relationship between range and power is more straightforward. By proper dimensioning, one can specify a desired range for the device and reduce the RF transmitting power if shorter distances are acceptable. In fact, dynamic power management for Wi-Fi devices has been proposed as a means of dealing with power consumption. Today, cellphones operate at significantly less power than when they were introduced in the early 1980s. Cellular systems dynamically reduce the output of handsets. It would be desirable if the Wi-Fi device could employ some form of dynamic power management. When the client is close to an AP and the signal is strong, it does not need as much power as when it is in a poor coverage area. Alternatively, in certain applications, it is possible to remove or bypass the RF system's PA altogether. Doing so not only reduces transmitting power consumption, but also limits the supported data rate to 1 or 2 Mbit/s. Research conducted by TI points to an interesting factor in determining the power consumption in Wi-Fi devices. It is logical to assume that higher data rates will result in higher power consumption, because higher transmitting power is needed to overcome interference and maintain the same range than with lower data rates. TI points out however, that the transmit duration, not the transmitting power, is the decisive factor.

Interestingly, transmitting or receiving at speeds of 11 Mbit/s or greater requires less power than at slower rates for a given amount of data. In addition, lowering the output power by removing or bypassing the PA (and reducing the range) has a measurable, but not significant, effect on the power consumed. Thus, embedded designs will incur only a small penalty in battery life by using a full-power, full-range radio.

An examination of the power budget for a PDA with 802.11 reveals that the incremental power consumption and the resulting battery life are quite reasonable. Based on simulation results of a Wi-Fi enabled PDA, TI estimated that more than 80 percent of the time, the Wi-Fi-enabled handheld was in a sleep or standby mode. The handheld spent the remaining 20 percent of the time in transmit or receive mode, which can be optimized with a higher data rate. Assuming the use of a 3,600-mW-hour Li-ion battery (1,000 mA hours) and an average daily power consumption of 105 mW hours, TI obtained a baseline battery life of about 34 days. With such results, adding Wi-Fi capability increases the power budget to approximately 140 mW hours per day. This provides an expected battery life of more than 25 days, a net reduction in power consumption of 25 percent, which makes it very reasonable for enabling wireless LAN connectivity.

8.8   Summary

We have covered a broad range of topics in this chapter. The main focus was on emerging trends in both technology and business practices. As Wi-Fi becomes more commoditized, we will see even more stringent criteria being applied to business decisions. The market is highly competitive, albeit attractive. Companies that wish to penetrate this market can be highly successful if they are laser-focused in their strategy and can clearly execute on their value proposition in this overcrowded market.

References

[1] Proposed Wireless Interworking Group (WIG) Baseline Document, Version 0.1, IEEE 802.11-03/004r0.

[2] IEEE 802.11 Submission for WNG SC, 11-02-294r1-WNG-HDR_802.11a_solution_using_MIMO-OFDM.ppt.

Glossary

3G – Third Generation

A term commonly referred to as an enhancement of voice-based second generation (2G) cellphones. 3G cellphones were expected to service a broad range of multimedia traffic.

3GPP – 3rd Generation Partnership Project

Responsible for the specification of UMTS. It covers the specifications of the UMTS Radio Access Network and the core network. A similar group (3GPP2) exists for CDMA2000 specifications.

AAA – Authentication, Authorization, and Accounting

AAA is the equipment that verifies user access to a network. Authentication verifies the user identity, Authorization verifies what the user is allowed to do, and Accounting bills for all the above according to different criteria such as time, data volume, application used, etc.

ACK – Acknowledgment

A short response packet sent by a receiving client to the transmitting client to indicate successful reception of a previously transmitted data packet.

AES – Advanced Encryption Standard

An encryption algorithm used by U.S. government agencies for unclassified material. It will eventually become the de facto encryption standard for commercial transactions in the private sector.

ARQ – Automatic Repeat Request

An error correction method in which the receiver employs error detection codes (e.g., cyclic redundancy check) to detect errors in the received packets and then request the transmitter to resend any corrupted packet. Compare FEC.

BSS – Basic Service Set

Essentially a wireless subnet (wireless coverage area) in which an access point services a group of wireless clients.

CCK – Complementary Code Keying

A coding method employed by the 802.11b extension for physical transmission. CCK increases the maximum data rate of the original 802.11 standard from 2 Mbit/s to 11 Mbit/s.

CDMA – Code Division Multiple Access

A spread-spectrum-based multiple access method in which each user is assigned a unique code (a spreading code) for identification and access to a common channel. Although the 802.11b extension also employs spread spectrum technology, it either uses a single spreading code or multiple codes to increase the data rate.

CDR – Call Detail Record

CDR contains detailed information about a connection and is used to bill telecommunications charges to the appropriate parties.

CSMA – Carrier Sense Multiple Access

A multiple access protocol in which a user first listens (senses) the channel to detect any ongoing transmissions before it attempts to transmit. If the channel is clear, the user transmits immediately, otherwise it waits for the current transmission to be completed before attempting transmission.

CSMA/CA – CSMA with Collision Avoidance

A CSMA variation adopted as a fundamental access mechanism (MAC protocol) by the IEEE 802.11 family of standards. The CA mechanism mandates that a user back off (delay transmission) for a random period of time immediately after a successful transmission. CSMA/CA is also known as DCF in 802.11 terminology.

DCF – Distributed Coordination Function

Same as CSMA/CA.

DES – Data Encryption Standard

A secret key cryptographic scheme standardized by NIST. Another popular encryption standard, Triple DES (3DES), is based on three successive invocations of DES.

DFS – Dynamic Frequency Selection

A dynamic function that changes frequency channels in response to radio interference.

DHCP – Dynamic Host Configuration Protocol

A protocol that allows a computer to be assigned an IP address automatically without manual configuration at a local server.

Diffserv – Differentiated Services

Any traffic management or bandwidth control mechanism that treats different traffic flows differently. A short tag is appended to each packet depending on its service class.

DIFS – DCF Interframe Space

The longest time interval used for transmitting 802.11 data packets. A longer time interval corresponds to a lower transmission priority. See also SIFS.

EAP – Extensible Authentication Protocol

An authentication framework that supports multiple authentication methods (EAP-MD5, LEAP, EAP-TLS, EAP-TTLS, EAP-SIM). IEEE 802.1X specifies how EAP can be encapsulated in LAN packets (EAP over LANs). Using EAP, the access point is blocked until the user is identified by the home AAA server.

FCC – Federal Communications Commission

A U.S. government agency responsible for the allocation of the radio spectrum.

FEC – Forward Error Correction

A coding mechanism that uses error correcting codes (e.g., convolutional coding) to detect with high probability the error location, thereby allowing the receiver to correct packet errors without requesting retransmission from the sender. Compare ARQ.

FFT – Fast Fourier Transform

An algorithm that allows individual subchannels to maintain their orthogonality (or separation) from adjacent subchannels. This technique allows received data to be reliably extracted and multiple subchannels to overlap in the frequency domain for increased spectral efficiency.

FIPS – Federal Information Processing Standard

One of a series of U.S. government documents that specifies standards for various aspects of data processing, including the Data Encryption Standard (DES).

GPRS – General Packet Radio Service

Often referred to as 2.5G cellphone technology, GPRS enhances existing GSM services such as circuit-switched data connections by providing end-to-end packet-switched data service at a user rate of around 40 Kbit/s.

GSM – Global System Mobile

GSM is the world's most widely used digital cellphone system. GSM standardization comes under the responsibility of the 3GPP.

HCF – Hybrid Coordination Function

A MAC protocol adopted by the 802.11g extension. The protocol allows the use of a deterministic, contention-free mechanism for arbitrating multiple transmitting users over a common wireless channel and has priority over the native DCF protocol. The removing of contention allows HCF to service time-sensitive traffic better than DCF.

HFC – Hybrid Fiber Coax

An improved cable television infrastructure that uses optical fiber for the backbone to interconnect coaxial cable feeds to the subscriber and allows two-way digital transmission in addition to television signals. Over 70% of all U.S. homes have been upgraded to the HFC architecture.

HiperLAN – High-Performance Radio LAN
A suite of wireless LAN standards developed by the European Telecommunications and Standards Institute (ETSI). The latest standard, HiperLAN Type 2 (or HiperLAN 2), is very similar to the 802.11a extension but not compatible.

HSO – Hotspot Operator
A company that runs a hotspot public wireless LAN service.

HTTP – Hyper-Text Transfer Protocol
The protocol used to download a Web page from one computer to another.

HTTPS – HTTP Secure
A popular method for ensuring secure access using a Web browser.

IAPP – Inter-Access Point Protocol
A protocol that facilitates roaming across access points from multiple vendors. Standardized under the IEEE 802.11f Task Group.

IDC – International Data Corporation
A popular company that provides technology forecasts, industry analysis, market data, and strategic guidance to builders, providers, and users of information technology.

IEC – International Electrotechnical Commission
A nonprofit organization dedicated to catalyzing positive change in the information industry and its university communities.

IEEE – Institute of Electrical and Electronic Engineers
A technical professional society promoting the development of electrical and electronic engineering. IEEE fosters the development of standards that often become international standards.

IETF – Internet Engineering Task Force
A technical professional society promoting the development of the Internet. IETF fosters the development of standards with the publication of Request for Comments (RFC) documents.

IFS – Interframe Space
A fixed time interval defined by the 802.11 standard to prioritize the transmission of control and data packets.

Intserv – Integrated Services
IntServ allows the network to support multimedia traffic using IETF's Service Level Agreements (SLAs) that provide explicit parameterization of traffic on a per-flow or per-packet basis, which corresponds to a certain level of service.

IP – Internet Protocol

This is a best-effort protocol in that it does not guarantee delivery of data packets.

IPSec – IP Security

An end-to-end security mechanism that runs on top of the underlying wireless or wired link. The mechanism employs a tunnel that encapsulates a packet in a higher-layer protocol for transmission.

ISM – Industrial, Scientific, and Medical

The ISM frequency bands are unlicensed. Created by the FCC in 1985, these bands sparked tremendous activities in commercial wireless LAN development.

ISO – International Standards Organization

An international standards organization best known for proposing the seven-layer Open System Interconnect (OSI) reference model for computer networking.

ISP – Internet Service Provider

A commercial company that offers connections to the Internet (e.g., telephone and cable companies).

Kerberos

A DES-based authentication system first developed by the Massachusetts Institute of Technology (MIT).

LAN – Local Area Network

A network that connects computing and peripheral devices within a limited geographical area, typically in a building. The IEEE 802 standards define a LAN to have a data rate of at least 1 Mbit/s.

MAC – Medium Access Control

The MAC protocol arbitrates the transmissions from multiple users over a single channel (e.g., a broadcast wireless channel). Also known as multiple access protocols or, just simply, access protocols.

NAT – Network Address Translation

A translation mechanism that provides connectivity for many computers using a single IP address, typically for the home. NAT changes the internal local IP address of each outgoing packet to a global, public IP address and vice-versa. This is done by rewriting the header of each incoming and outgoing packet. An extension known as Network Address and Port Translation (NAPT) allows multiple nodes to use one IP address on a public network. In this case, Layer 4 ports are used to distinguish the different nodes.

NIC – Network Interface Card

A hardware device that plugs into the computer and connects the computer to a network.

NIST – National Institute of Standards and Technology
An agency of the U.S. government. Formerly known as the National Bureau of Standards.

OFDM – Orthogonal Frequency Division Multiplexing
A multiplexing technique that forms the basis of the 802.11a and 802.11g extensions. The method involves combining many radio subchannels (or subcarriers), each transporting a portion of the information contained in a data packet.

OSI – Open System Interconnection
An international standard that introduced the layered concept in networking design.

PBCC – Packet Binary Convolutional Coding
An optional coding method employed by the 802.11b and 802.11g extensions for physical transmission. PBCC increases the maximum data rate of the CCK coding scheme from 11 Mbit/s (in 802.11b) to 22 Mbit/s (in 802.11b) and 33 Mbit/s (in 802.11g).

PCF – Point Coordination Function
An optional multiple access mechanism (MAC protocol) adopted by the IEEE 802.11 family of standards. The mechanism eliminates contention on a broadcast wireless link by allowing an access point to poll each individual user for data. This in contrast to the DCF method, in which each user, including the access point, makes its own decision to transmit.

PCMCIA – Personal Computer Memory Card International Association
An association best known for creating the PC card standard.

PDA – Personal Digital Assistant
A hand-held device that provides computing and information storage.

QoS – Quality of Service
With QoS, service differentiation is possible and statistical guarantees (e.g., data rates, error rates, delay jitter) can be accorded to applications that require them.

RADIUS – Remote Access Dial-In User Service
A service that provides mutual authentication between the client and the authentication server.

RSVP – Resource Reservation Protocol
A reservation protocol that assumes QoS options are needed but not as a replacement for best effort service.

SDR – Software Defined Radio
A technology that allows different wireless standards to run over a common hardware.

SIFS – Short IFS

The shortest time interval that is used for higher-priority 802.11 control packets such as acknowledgments.

SIM – Security Identity Module

An electronic card that is normally used to securely verify the identity of a cellphone user. Without the SIM card, a cellphone is virtually useless.

SLA – Service Level Agreement

Refers to the QoS contracts. The usual agreement specifies the end-to-end performance to which the client is entitled over a specified period of time.

SMTP – Simple Mail Transfer Protocol

The protocol used to send e-mail from one computer to another over the Internet. SMTP is part of the TCP/IP protocol suite and is standardized by the IETF.

SNMP – Simple Network Management Protocol

The protocol that specifies, using standardized management information base (MIB) messages, how a network management station can communicate with remote devices such as routers. To monitor or control a remote computer, a manager must fetch or store values to MIB variables. SNMP is standardized by the IETF.

SOHO – Small Office, Home Office

A term commonly used to refer to a small enterprise.

SSID – Service Set Identifier

The SSID is a common network name for devices operating in the same wireless subnet served by an access point. The SSID is not well protected since it is broadcast by the access point to all users in the wireless subnet.

SSL – Secure Sockets Layer

A mechanism invented by Netscape Inc. to provide secure communication between a browser and a server.

TCP – Transmission Control Protocol

The transport layer protocol that provides application programs with access to reliable, connection-oriented service. TCP offers reliable delivery and adapts to changing traffic conditions on the Internet by controlling the rate at which a sender transmits data.

TCP/IP

Two of the most important protocols in Internet's protocol suite.

TKIP – Temporal Key Integrity Protocol

Initially referred to as WEP2, it uses the same encryption as WEP, but changes temporal keys regularly. Existing equipment can be upgraded through simple firmware patches and WEP-only equipment can still interoperate with TKIP-enabled devices.

TPC – Transmit Power Control
Allows a dynamic response to radio interference by using lower power modulation.

UDP – User Datagram Protocol
A transport layer protocol that provides application programs with connectionless communication service, as opposed to the connected-oriented service provided by TCP.

UMTS – Universal Mobile Telecommunications System
One of the two main systems for 3G cellphone systems recognized by International Telecommunication Union (ITU). UMTS is composed of two different, but related, modes: FDD (Frequency Division Duplex) and TDD (Time Division Duplex). FDD mode is considered as the main technology; it supports data transfer rates up to 384 Kbit/s.

VPN – Virtual Private Network
A private network constructed over a public network that employs encrypted IP tunneling to allow a multisite enterprise to communicate securely over the Internet.

WAN – Wide Area Network
A network that spans large geographic areas, typically on a nationally basis. Due to the larger coverage area, WANs have higher propagation delay than LANs.

WAP – Wireless Application Protocol
A wireless Web protocol targeted for cellphones.

WECA - Wireless Ethernet Compatibility Alliance
The group behind the testing and certification for interoperability of 802.11-based networking products. It has recently changed its name to the Wi-Fi Alliance.

WEP – Wired Equivalent Privacy
A security protocol that encrypts a data packet before it is transmitted to/from the access point.

WFA – Wi-Fi Alliance
See WECA.

Wi-Fi – Wireless Fidelity
A stamp of approval for 802.11b products certified by the Wi-Fi Alliance.

WPA - Wi-Fi Protected Access
WPA is based on the portions of the 802.11i extension that can be implemented on software.

WISP – Wireless ISP
A company that provides connectivity to the Internet, mostly in hotspot public wireless LAN.

WISPr – WISP Roaming

An agreement among hotspot service providers that allows users to roam from one WISP to another and yet have billing information consolidated.

xDSL – x Digital Subscriber Line

A generic acronym that refers to any DSL local loop technology such as ADSL, HDSL, VDSL, etc. These technologies supply high-speed broadband data rates over the same twisted pair copper wiring used for traditional telephone service.

Related Web Sites

Standards

3rd Generation Partnership Project
www.3gpp.org

3rd Generation Partnership Project 2
www.3gpp2.org

IEEE 802.11 Working Group for Wireless Local Area
Networks
http://ieee802.org/11

IEEE 802.15 Working Group for Wireless Personal Area
Networks
http://ieee802.org/15

IEEE Wireless Standards Zone
http://standards.ieee.org/wireless

ETSI Standards
www.etsi.org/getastandard/home.htm

Bluetooth
www.bluetooth.com

Wi-Fi Alliance
www.wi-fi.org

Wireless LAN IC Vendors

Advanced Micro Devices (AMD)
www.amd.com

Agere Systems
www.agere.com

Atheros Communications
www.atheros.com

Broadcom Corporation
www.broadcom.com

Intel Corporation
www.intel.com

Intersil Corp
www.intersil.com

Marvell Communications
www.marvell.com

Philips Semiconductors
www.philips.com

RF Micro Devices (RFMD)
www.rfmd.com

Texas Instruments
www.ti.com

Index

About the Authors

Teik-Kheong (TK) Tan (tktan@ieee.org) is Chairman of the Wireless LANs Next Generation Committee of the IEEE 802.11 Standards Working Group. He also co-chairs the Wireless Interworking Group, a committee formed jointly by IEEE 802.11, ETSI BRAN, and MMAC. He was formerly co-chair of the Wireless Ethernet Compatibility Alliance (WECA) Marketing committee. As a technology evangelist and product specialist for various companies (most recently, Philips), TK has over 15 years of managing strategic alliances and standards/regulatory experience in both broadband and wireless industries. On the wired front, TK is also actively involved in broadband networking. He was the Vice President of The ATM Forum and an elected member of the ATM Forum Board of Directors with main responsibilities in technical, marketing and promotional aspects of ATM in the Asia-Pacific region. A frequent speaker and panelist in many international and regional industry conferences, TK was also the President and founder of the ATM Special Interest Group (Singapore) and served as Chairman of the Singapore IEEE Communications Chapter (1995-1996). He served as Editor-in-Chief of the IEEE Communications Society Asia Pacific Newsletter from 1997-1999. He has published numerous papers in international Journals and conferences, and was the IEEE Reviewer on Broadband Networks, Internet and Communications Systems. He has also received numerous international awards for his worldwide contribution to networking and is listed in the International Who's Who of Professionals as an honorary member.

Benny Bing (bennybing@ieee.org) is a research faculty member with the School of Electrical and Computer Engineering at the Georgia Institute of Technology. He has published over 30 research papers and six books, including *Wireless Local Area Networks* (an abridged version of his first book titled *High-Speed Wireless ATM and LANs*), which has been adopted by Cisco Systems worldwide. His publications have also appeared in the *IEEE Spectrum*. He is currently a technical editor for *IEEE Wireless Communications* and has also guest edited for *IEEE Communications Magazine* and *IEEE Journal on Selected Areas on Communications*. Recently, he was featured in *MIT Technology Review* in a special issue on wired and wireless technologies. He consults for industry and is a frequent lecturer on wireless LAN subjects, having conducted customized on-site courses for Qualcomm Inc. in San Diego, CA and the Office of Information Technology at Georgia Tech. He serves on the wireless networking panels for National Science Foundation and Building Industry Consulting Services International (BICSI). In addition, he currently serves on the technical program committees of several IEEE conferences and chairs the IEEE International Conference on Wireless LANs and Home Networks (www.icwlhn.org). He is a recipient of a best paper award at the 1998 IEEE International Conference on ATM and the Lockheed-Martin Fellowship. His current research interests include wireless local area networks, broadband first-mile access, protocol design, and queueing theory.